Water Resources Series
Volume 5

Rain and Stormwater Harvesting in Rural Areas

WATER RESOURCES SERIES

Volume 1: WATER RESOURCES MANAGEMENT IN INDUSTRIAL AREAS
Selected Papers from the International Symposium held by the International Water Resources Association (IWRA), 7-11 September 1981, Lisbon, Portugal.

Editors: Leo R. Beard and W.H.C. Maxwell.

Published for the International Water Resources Association.

Volume 2: WATER FOR HUMAN CONSUMPTION: Man and his Environment (Volume I)

A selection of papers prepared for the IVth World Congress of the International Water Resources Association (IWRA), 5-9 September 1982, Buenos Aires, Argentina.

Published for the International Water Resources Association.

Volume 3: LONG-DISTANCE WATER TRANSFER: A Chinese Case Study and International Experiences

Editors: Asit K. Biswas, Zuo Dakang, James E. Nickum and Liu Changming.

Published for the United Nations University.

Volume 4: RIVER BASIN DEVELOPMENT

Proceedings of the National Symposium on River Basin Development, 4-10 December 1981, Dacca, Bangladesh.

Editor-in-Chief: Munir Zaman; *Editors:* Asit K. Biswas, Amjad H. Khan and Ainun Nishat.

Published for the United Nations Development Programme.

Volume 5: RAIN AND STORMWATER HARVESTING IN RURAL AREAS:
A Report by the United Nations Environment Programme.

Volume 6: WATER FOR HUMAN CONSUMPTION: Man and his Environment (Volume II)

Proceedings of the IVth World Congress, International Water Resources Association (IWRA), 5-9 September 1982, Buenos Aires, Argentina.

Editor: W. Hall C. Maxwell.

Published for the International Water Resources Association

Rain and Stormwater Harvesting in Rural Areas

A Report by
The United Nations Environment Programme

Published for the UNITED NATIONS ENVIRONMENT PROGRAMME
by TYCOOLY INTERNATIONAL PUBLISHING LIMITED
DUBLIN

Published by:
Tycooly International Publishing Limited
6, Crofton Terrace,
Dun Laoghaire,
Co. Dublin, Ireland
Telephone: (+ 353-1) 800245, 800246
Telex: 90635 TYCO EI

First Edition 1983

Phototypeset by AfricaScience International Publishing Ltd., Nairobi, Kenya
Printed in Ireland by Irish Elsevier Printers Ltd., Shannon.
Cover design by Alanna Corballis
ISBN 0 907567 38 X Hardback.
ISBN 0 907567 39 8 Softcover.

CONTENTS

Introduction

THIS BOOK is based on discussions by consultants of the United Nations Environment Programme who met in Nairobi on 30th October 1979 to discuss reports on rain and stormwater harvesting in Africa, Asia, Australia and the Pacific, Central America, China and the Middle East.

Water harvesting refers to the deliberate collection of rainwater from a surface (catchment) and its storage to provide a supply of water. This process is distinct from the natural runoff of water into perennial rivers which is then controlled and stored in dams and reservoirs. There are many regions in the world where rainfall is heavy for some months of the year and light for the rest; rainwater and storm runoff, harvested in season and then stored, would help in alleviating the problem of water shortage during the dry season.

The purpose of the meeting in Nairobi was to bring together the various technologies for water harvesting and to select feasible and effective methods for follow-up demonstrations in the field, with the operational aim of assisting communities in climatically suitable areas to undertake the harvesting of their own water supplies. The information presented here by no means exhausts existing knowledge and the recommendations made represent the first efforts to focus attention on several methodologies to promote a wider application of safe water harvesting practices. Thus, a summary of world-wide technologies has been compiled relating to the collection, storage, treatment, distribution and use of rainwater and other forms of precipitation.

CHAPTER 1

Rainwater Harvesting for Domestic Use—Early History

MUCH of the early history of rainwater harvesting has its origins in Europe, where the location of towns and industrial and commercial centres was determined by physico-geographical factors, the principal factor, of course, being the availability of water for domestic and communal use, and later for industrial consumption.

Water management has a long tradition in Europe. The first reservoirs were constructed in the Roman Empire (figures 1-1A and B). Generally, small earth and masonry dams were built not only to store runoff water but also to raise the general water level. The Romans achieved considerable control of runoff by building structures to modify the outflow of lakes. For example, the first structure to control the outflow of Lake Balaton in the Roman province of Pannonia—now Hungary—was built in the era of Emperor Galerius between 305 and 311 AD.

The size and structure of dams did not change in the centuries following the fall of the Roman Empire. The water demand was, however, still relatively modest and large reservoirs were not required. Runoff was collected for the primary purpose of running watermills, forming fish ponds or for recreational purposes. The Romans even used runoff for defensive purposes such as surrounding forts with strategic lakes and marshes.

The rapid increase of reservoirs both in size and in number began only a hundred years ago. This change was associated with the use of new building materials, including concrete and steel structures, and the implementation of new designs, such as arch dams.

There is, however, a region in Europe, which was exceptional from the point of view of runoff control—the islands and coastal areas of the Mediterranean. The climatic conditions were unfavourable from the point of view of water resources development. To cover the water demand, methods simpler than reservoir construction were needed for application under conditions of low precipitation and high evaporation. These methods included the collection and storage of rainwater, utilizing roofs, courtyards and other microcatchments.

The amount of drinking, cooking and washing water required in ancient European civilizations was relatively small and the areas for collecting the water by rain harvesting were incorporated into the design of houses, for example, roofs and paved courts. Basins or cisterns were also constructed to store water for domestic purposes, and drainage systems connecting the collecting surfaces with storage facilities were common.

Figure 1-1A A masonry dam in Pannonia province

Figure 1-1B The spillway of the dam shown above

Figure 1-2A Reconstructed corner of an atrium in Aquincum Museum, Budapest, Hungary

The development of this technique actually dates back to prehistoric times. In the ruins of the palace of Knossos (1700 BC), the centre of Minoan Crete, sophisticated examples of rain harvesting have been found. Within the blocks of both the western and eastern wings of the palace, lighting shafts or small courts were formed between the rooms. Apart from providing light for the lower floors, their purpose was also to collect rain water. This water was drained out of the palace from the residential quarter. In the Court of the Stone Spout, the rainwater was collected from the roof of the Great East Hall through a stone drain. The water was channelled to a small cistern and then through a pipe to a blind well. Near the northern entrance six oblong cisterns (Minos' Cells) about eight metres deep were found with traces of hard plaster lining, indicating that the cells might have been used for water storage, although the water collecting systems have not been discovered. The throne room also provides an example of another application of rain harvesting. Facing the throne is a basin at a lower level below the lightingshaft of the chamber. The rainwater collected here might have served for ritual activities, such as a ritual of purification before approaching the king, but most probably its main purpose was the air-conditioning of the room.

The design of the typical form of Roman house indicates that the domestic water supply was based on the rainwater harvesting technique used in Minoa. The ground plan of the house is oblong with the entrance at one of the shortest sides. Through a small anteroom (vestibulum) a closed internal court is reached (atrium) which has a pool in the centre (figure 1-2A) into

Figure 1-2B Stone gutter in the centre of a paved court in Aquincum, which collects and leads water into the cistern located below the court

which the rainwater from the roof flows. The pool is lined inside by ceramics and has sloping sides. The atrium is closed behind by the sitting room (tablinium) which leads into the larger internal court (peristilium) where a second pool collects water from the roofs of the rooms surrounding the court. Originally the pools had a double purpose, storing water needed for domestic use and also the air-conditioning of the courts. Later, when the number of inhabitants increased, and the extension of grounds was restricted because of the more intense use of building land for expanding towns, it was necessary to increase the storage space. The protection of the water against pollution as well as the lowering of evaporation losses then became necessary. Deeper, larger closed cisterns were constructed below the court and the greatest part of the water was drained from the roofs through sieves (figure 1-2B) and fed into the cisterns. The shallow pools then served only for air-conditioning. With the development of large cities, central water supply systems were established to meet domestic demands, and rain harvesting lost its importance. Today only the pools in the courts remain as a reminder of this earlier form of water supply. Later even the water in the pools was piped from the central system in order to create a fountain and thus enhance the air-conditioning effect of the pool.

The double purpose of the basins in Roman houses produced two different developments in medieval architecture. Most of the early European

Figure 1-3 The cistern of Fort Csesznek, Hungary, showing the niche for the filter on the left and the tunnel for transporting the water on the right

fortresses were built on rocky peaks. The domestic water for the inhabitants was usually carried up from the valleys. But when this supply was cut off during sieges rainwater was collected from the roofs by means of gutters and channelled to cisterns to create a reserve water supply. The cisterns generally consisted of blind wells lined with cut-stone. The masonry was made watertight by clay backing and by setting the stones and bricks in clay.

In order to protect the water against pollution, the store of water was filtered through a gravel layer before flowing into the cistern. In figure 1-3, for example, the niche to the left of the cistern housed the filter.

Another application of rain harvesting can be seen in the Mediterranean region, especially in Spain, where air-conditioning of closed internal courts is provided by collecting rain from surrounding buildings. At a later stage in this development a more stable water supply succeeded rain harvesting, and this advance made the construction of fountains possible. One of the best known examples of this application of basins in courts for cooling is in the Alhambra Palace in Granada (figure 1-4).

Figure 1-4 The Arrayanes Courtyard of Alhambra, Granada, Spain

At the beginning of this century rain harvesting lost its importance because the water demand was much higher than could be provided for by such a simple method. However, there are small areas (especially in regions where only a highly mineralized groundwater is available for domestic use) where the inhabitants still utilize the rain harvesting technique as a supplementary water source. Water from the roofs is collected by rainwater gutters and fed into concrete or steel tanks and used for washing and cooking, since the high salt content of the groundwater renders it unusable for such functions. At the present time in Europe, the application of rain harvesting for domestic water supply has been largely superseded by the introduction of centralized piped water supply. This applies even in remote villages.

CHAPTER 2

Rainwater Harvesting for Agriculture—Early History

RAIN and stormwater harvesting techniques for agriculture were extensively practised throughout a vast region of North America, through the mountains and basins of the Mexican northwest, and through the civilizations which flourished in the south central highlands of Mexico. These techniques were also practised throughout the Middle East, North Africa, China and ancient India.

Although the impermanent quality of some of these structures does not allow a precise assessment of their antiquity, current archaeological evidence suggests that rainwater harvesting was widespread by the ninth or tenth centuries AD. Because of these studies, we now have a good picture of water harvesting in southeastern USA and nothern Central America. Many examples can be chosen from this region to illustrate water harvesting.

These early techniques made it possible to cultivate arid and semi-arid areas which otherwise might not have supported farming or permanent settlements. Many investigators argue that, for this reason, water harvesting was pivotal in sustaining the high population densities associated with the evolution of Meso-American civilization. These authorities, moreover, point to another key feature of ancient water harvesting, namely, its close association with the key staples of pre-Columbian diet: maize, beans, squash and a drink made from agave (*maguey*).

It is important to note, however, that water harvesting was only one of a broad spectrum of moisture management techniques (ranging from dry farming to large-scale irrigation) for supporting sedentary life. In addition, especially in the more arid zones, farmers depended heavily on hunting and the gathering of food to supplement their crop harvests.

Since water harvesting was integrated within a flexible, multiple-option strategy of resource use, in most cases it becomes impossible to determine its precise contribution to overall livelihood support. Nevertheless, from the widespread distribution of relics of water harvesting structures as well as their persistence through time, it is reasonable to assume that water harvesting techniques played an important role in the successful settlement of North America's vast drylands and also in other present-day desert areas, such as the Negev.

Life in pre-Conquest North America centred on an agricultural economy based on maize, beans, squash, chillis and agave maguey. Water harvesting on hillslopes and hillside valleys played an important role in this economy, although the magnitude of this role is difficult to determine. However, it is possible to draw some preliminary inferences from a comparison of the settlement patterns in the moister southern regions with those of the arid northern frontier.

Archaeological records show that agriculture supported a wide variety of settlements, from dispersed hamlets to fully developed urban centres. The dense populations and cities of central Mexico were probably the result of favourable rainfall conditions permitting a wider range of moisture management strategies to come into play, including irrigation and intensive lakeshore cultivation.

Further north, under more severe arid conditions, we find that the range of strategies was restricted. It is probable that water harvesting became a prime moisture management technique in the American southwest, where we find permanent sedentary agricultural populations who developed a complex social organization and a sophisticated artistic and ritual life, but who never attained full urbanization.

Although it is difficult to specify the numbers of pre-Conquest populations that were supported directly or exclusively by water harvesting, it is possible to give some indication of the spatial extent of various harvesting technologies. For example, contour terracing, a common rainwater harvesting technique, has been documented at archaeological sites throughout the American southwest and the Mexican northwest. Although the ephemeral nature of these structures makes it difficult to document the existence of contour terraces unambiguously, few experts would doubt their widespread existence prior to the Conquest.

Stormwater harvesting techiques using silt traps and check dams are also thought to have been widespread. Somewhat less ephemeral than contour terraces, the remains of check dams have been discovered at numerous sites throughout the central highlands of Mexico and the American southwest.

Over the past two decades, archaeological investigations have yielded a wealth of information on water harvesting and other water management practices in pre-Columbian North America. Yet the archaeological record provides us with only partial knowledge regarding the nature and distribution of the numerous water-harvesting techniques practised by ancient cultivators. It is probably safe to argue that water harvesting was even more extensively practised than the current record suggests.

This present examination of ancient water harvesting reveals that it possessed two features. First, it was extremely flexible; and second, it was remarkably enduring. Its flexibility is demonstrated in terms of its easy integration with other resource-use systems, as well as by its widespread adoption by diverse cultural groups in various parts of the world. Its enduring qualities are reflected in its antiquity and its capacity to persist in the face of abrupt changes in the social order.

There is also another characteristic feature of water harvesting which has emerged with equal force over the centuries: its association with peoples inhabiting marginal environments. We find that water harvesting has survived the Spanish Conquest, the introduction of new crop varieties and livestock, the agrarian reform, the green revolution, and is still widely practised by Mexican farmers on lands that other groups consider too marginal for their needs.

The labour requirements for most water harvesting techniques were modest, mostly within the capabilities of individual households or small communities. Unlike other technologies, water harvesting did not require a large-scale centralized power structure for its construction, operation and maintenance. Moreover, many water harvesting structures were constructed incrementally, frequently built over several decades or generations.

Based on inferences that can be made from the archaeological record, it seems clear that water harvesting made the difference between the presence and absence of both agriculture and settlements in large areas of North America which did not have permanent sources of water such as streams or springs which would have permitted irrigated cultivation.

The capacity of water harvesting to make the difference between effective agricultural occupancy and less intensive non-agricultural land-use has been widely recognized. However, an evaluation of the adequacy of water harvesting must recognize its vulnerability to seasonal fluctuations in rainfall.

The fact that water harvesting was not foolproof and could result in crop failures, must have impelled ancient cultivators to integrate their water harvesting practices with other water management techniques.

In the American southwest where water harvesting was the principal source of agricultural water, it was supplemented by water from occasional springs. Some irrigation was practised along the few perennial streams. In some wet upland areas, rainfall was adequate for rainfed farming.

In the central highlands of Mexico the rich ecological mosaic provided opportunities for a complex integration of moisture management alternatives. These included permanent rainfed farming, catch cropping, small- and large-scale irrigation, drained field agriculture, river bottom farming, intensive lake cultivation (*chinampas*) along with a full range of water harvesting technologies. Two types of rainwater harvesting—contour terracing and bordered gardens—were the most important for ancient agriculture.

Although other water harvesting techniques such as microcatchments, sand-dune farming, and floodwater farming are also of ancient origin, there remains little conclusive evidence of their former use. Their description will be included in later chapters dealing with contemporary water harvesting techniques.

The examples of ancient rainwater harvesting involve water and moisture control at a very simple level. Often they consist of nothing more elaborate than rows of rocks placed along the contours of slopes. This simplicity, along with the fact that very few of these structures can be unambiguously traced back to historic and prehistoric times, should not blind us to the possibility that simple rainwater harvesting techniques such as contour terracing were widely practised and possibly provided the mainstay of agricultural life for millions of people throughout sub-humid North America.

Contour terraces (also termed linear borders, terraces, semi-terraces, sloping terraces, *trincheras* and *metlepantli*) were constructed by placing long rows of stones spaced at intervals along the contours of a slope. Since these rows lay perpendicular to the gradient of the slope, they were designed simply to trap slope wash and therefore resulted in relatively minor changes in slope profile (figures 2-1 A and B, 2-2 and 2-3). Runoff captured behind these barriers also allowed for the retention of soil, thereby serving as an erosion control measure on gentle slopes. In contemporary examples of contour terracing these modest structures are frequently reinforced by earth embankments or economically useful plants such as the agave maguey. Such features have not been preserved in the archaeological examples.

Some of the most important sites with remains of contour terracing documented through archaeological surveys include Point of Pines (Arizona), Mesa Verde (Colorado), Chaco Canyon (New Mexico), and on the northern Rio Grande (New Mexico). In Mexico, ancient contour terraces have been surveyed in the Rio Gavilan (Chihuahua), the Tehuacan Valley (Puebla), and the Nochixtlan Valley (Oaxaca).

Contour terracing in the American southwest offers the best evidence of ancient contour terracing. Good examples come from an archaeological site known as Point of Pines located in east central Arizona. The topographic setting is one of low ridges and open valleys with typical vegetation including grasslands, Ponderosa pines, piñon and juniper. Soils are shallow and erode easily when the natural vegetation cover is removed. Average annual precipitation measures between 450 and 500 mm falling in short, intense summer rain or hail storms.

Point of Pines was occupied in 2000 BC by hunter-gatherers; in 100 AD small-scale agriculture had begun. Three types of water harvesting were in use after 1000 AD: contour terraces, check dams, and bordered gardens. A major survey of Point of Pines carried out in the 1950s identified ten sites totalling 75 acres scattered over 100 square miles; many similar sites remain unsurveyed.

Based on existing remains, contour terracing appears to have been the most common water harvesting technique used by Point of Pines cultivators. The usual practice consisted in placing rows of boulders or stones along slope contours. The dimensions of the rows varied from the size of a single stone to one metre wide, and the original height corresponded to that of two to three stones. Although most rows were laid in parallel lines along the gentler (5 per cent) slopes, some were arranged in concentric circles around knolls near house sites.

The largest site surveyed at Point of Pines included house sites and a group of 29 contour terraces built on 14 per cent slopes. The distance between stone rows ranged from one to five metres. As the slope became gentler (3 per cent), the terraces were spaced more widely, from 5 to 25 metres apart.

The structures appear to have been adequate for water and soil control. Even with the terraces in considerable disrepair, the soil today is still deeper on the upslope side of the stone rows.

Figure 2-1 Relic check dams, *trincheras,* in the Rio Gavilan Region: (**above**) long gully; (**below**) detail (Herold, 1970).

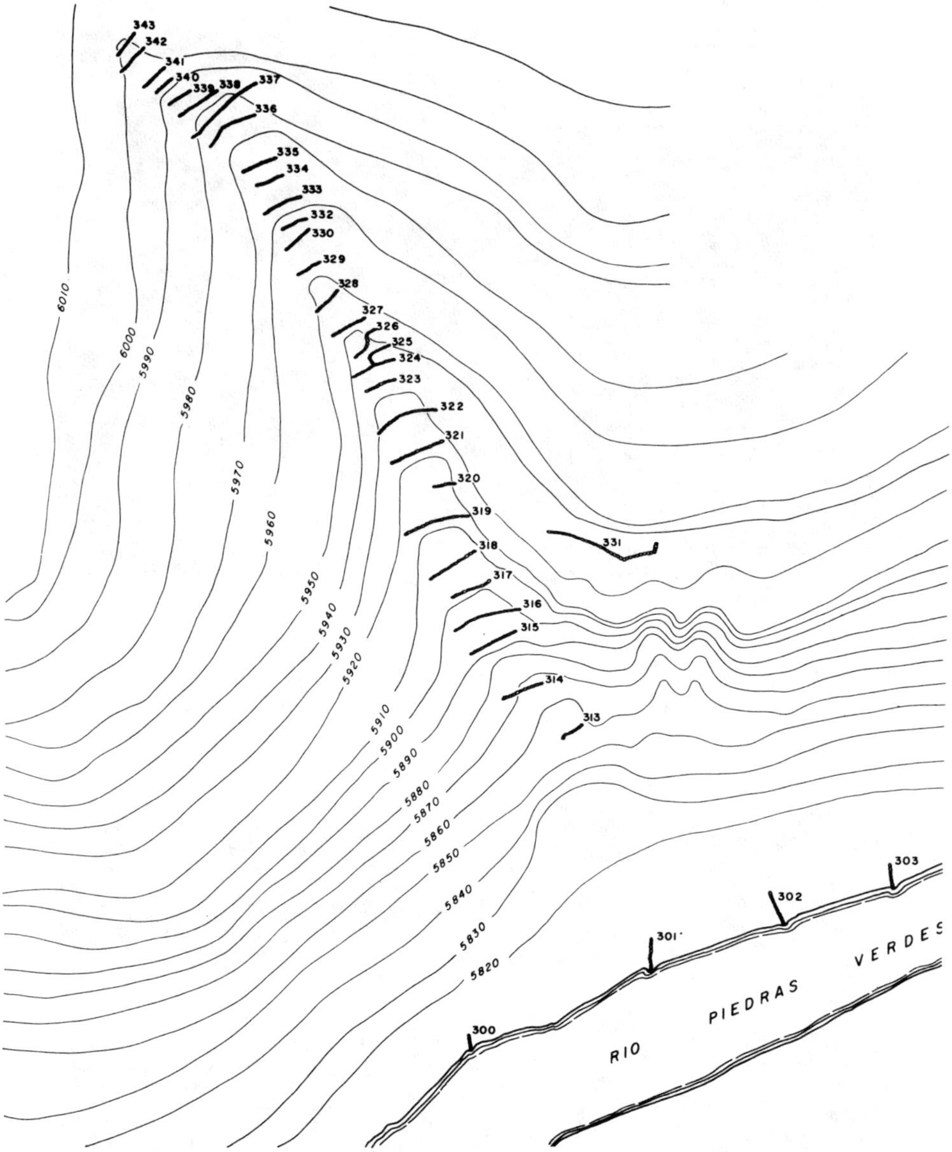

Figure 2-2 Relic check dams, *trincheras,* in the Rio Gavilan Region (Herold, 1970)

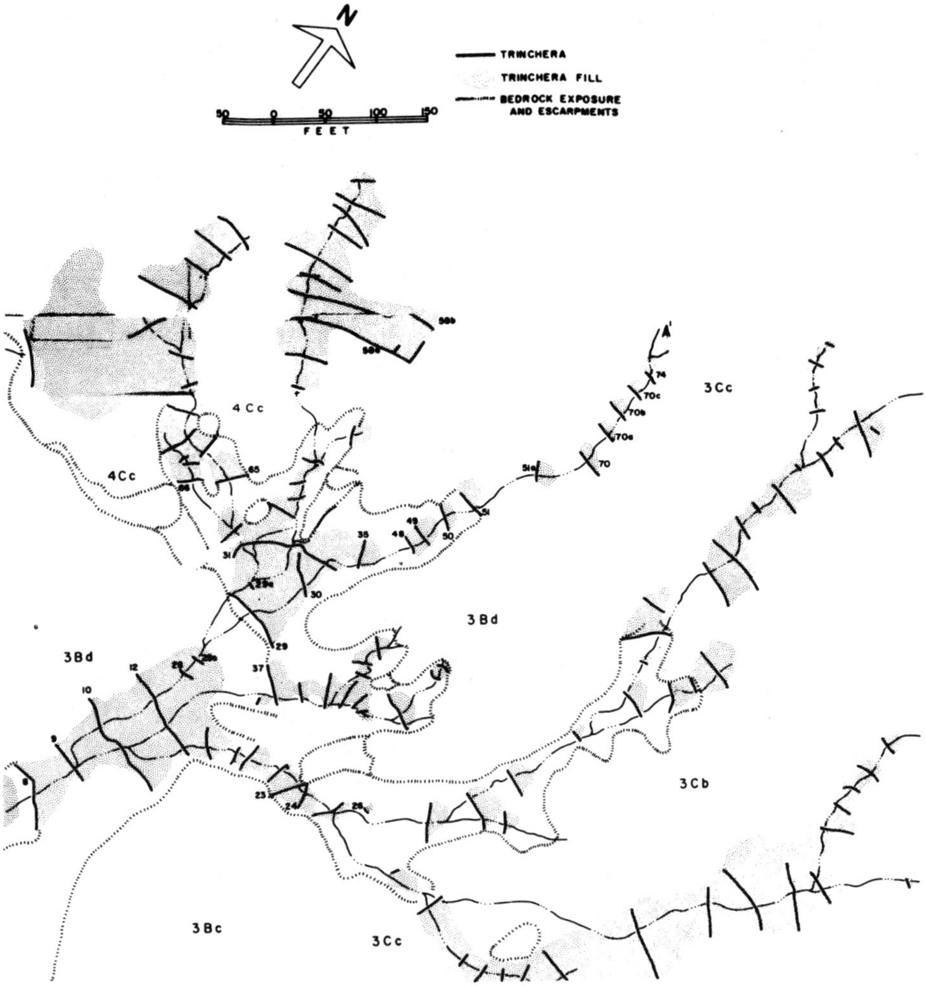

Figure 2-3 Relic check dams, *trincheras,* in the Rio Gavilan Region (Herold, 1970)

Researchers estimate that the total cultivated area of the surveyed portions of Point of Pines amounted to approximately 2,500 hectares. Of these, 15 per cent were supported by water harvesting structures. It is possible that a total of 3,000 people were supported by this agricultural base. Point of Pines was abandoned by its original inhabitants during the fifteenth century AD, one of many mysterious population constrictions that occurred throughout the southwest at that time.

The most extensively practised form of water harvesting in pre-Columbian and Colonial Mexico was contour terracing on gentle slopes in which slope runoff was trapped behind low stone structures, earth embankments, or hedges of agave 'maguey' plants placed in long rows at even intervals perpendicular to slope gradient. The series of parallel elongated fields formed in this fashion are known as *metlepantli, bordos* or *melgas*.

Metlepantli are thought to have contributed (along with other water management techniques) to the support of the astonishingly high population densities (up to 130 persons per square mile) reached in pre-Conquest Mexico.

The relationship between contour terracing and demographic change has been the subject of some speculation. It has been argued that extensive rainwater harvesting on gentle and medium slopes was a response to population pressures in pre-Conquest times. Moreover, the precipitous demographic decline of the Colonial period is thought to have led to the abandonment of hillslope terrace systems resulting in their deterioration and, consequently, severe erosion. If this interpretation of long-term trends is correct, it suggests that contour terracing can be highly sensitive to fluctuations in the availability of labour for maintenance and furthermore, that this sensitivity can have serious ecological consequences.

Systematic archaeological surveys of ancient contour terraces are hampered because the structures have either disappeared or because their continued agricultural use prevents definitive dating. However, remnant terraces have been identified in the Rio Gavilan region of the nothern Sierra Madre (see figure 2-4) and in the Tehuacan and Teotihuacan Valleys of Central Mexico. Much of our present knowledge regarding these systems is based on inference and speculation. Beyond the particulars of the physical structure of what remains, the archaeological record provides us with very little information on their social setting or effectiveness. The reader should turn to Chapter 7, which deals with contemporary contour terracing, for a full description of the technical and socio-economic aspects of this technology.

Although not as widely used as contour terracing, relic bordered gardens are found at many sites in the American southwest including Point of Pines, in New Mexico along the northern Rio Grande, in Chaco Canyon, and at the Rainbow Plateau along the Utah/Arizona border. Undoubtedly many examples exist in locations which remain unsurveyed. This technology is not known to have been practised in Mexico.

Bordered gardens are small areas of fertile soil surrounded on four sides by stone walls or earth ridges. They are fed by external sources of water such as seeps, springs or impounded runoff. This water is routed to the gardens by means of channels or ditches (figure 2-5). Perhaps the best example of relic bordered gardens is to be found at Chaco Canyon in northwestern New Mexico. The Chaco Canyon is located between 1,500 and 2,100 m in an area characterized by broad plains, mesas and shallow canyons. The higher elevations are covered by piñon and juniper. Rainfall averages between 125

Figure 2-4 Relic contour terraces at Rio Gavilan (Herold, 1970)

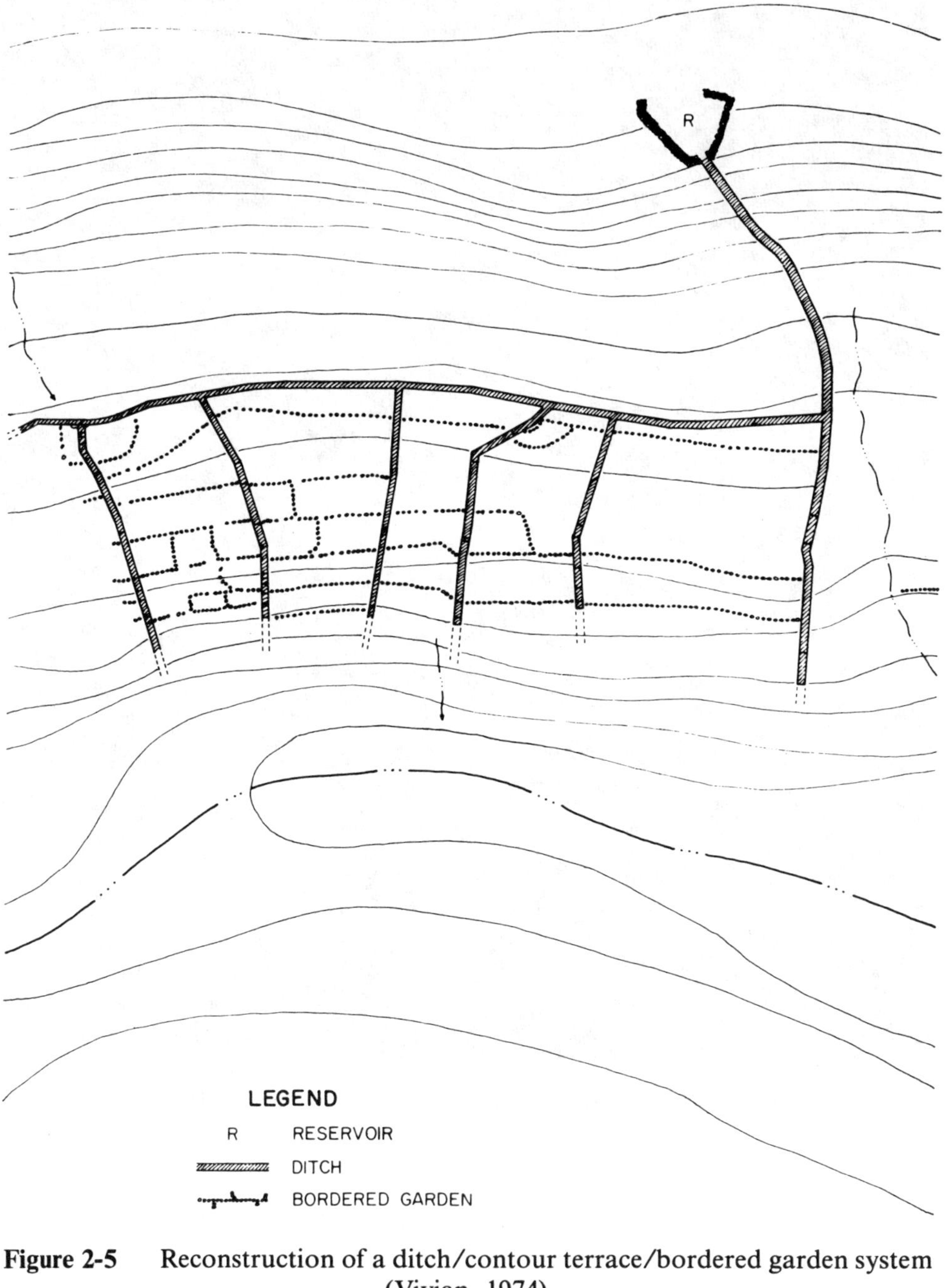

Figure 2-5 Reconstruction of a ditch/contour terrace/bordered garden system (Vivian, 1974)

and 175 mm a year; much of this falls in the form of short intense summer rainstorms.

The elaborate border garden systems of Chaco Canyon incorporated diversion dams, canals, ditches, headgates and earth-bunded fields. The gardens depended on harvesting runoff from 28 large and small drainage basins representing a total catchment of 4,250 hectares.

At one site (figure 2-6), water harvested by diversion dams from the slopes was directed by ditches to a canal (in places masonry lined) averaging 4.5 m wide and 1.4 m deep. The canal extended 230 m to a multiple headgate which slowed the water's flow and channelled it to the bordered garden complex.

The typical garden complex covered between 8 and 9 hectares and was divided by the canal into four large sections, each containing 84 bordered plots, averaging 322 square metres each. Each small garden plot was watered by canal waters flowing through temporary breaches in the garden borders.

The meagre annual rainfall, interspersed with years of intense, destructively high precipitation, required a water harvesting system which could efficiently regulate the water available to the garden plots. The adequacy of this system has been illustrated by measurements of a recent heavy rainstorm. The storm provided 300 mm of rain in a single hour. It was estimated that this would have provided a 9.7 hectare garden complex with 204,000 decalitres of water. However, this harvested water could only be of use if the diversion dams, ditches and headgates were designed in such a manner that the flow of the water could be modulated so as not to destroy the garden plots.

It is estimated that in 1050 AD up to 10,000 individual bordered gardens (5.26 to a hectare) fed a population of 10,000, who lived in small towns and hamlets concentrated in an area 14.5 km by 1 km in New Mexico.

In Europe rainwater harvesting began when small, crude retaining walls were constructed on slopes. Later the clearing of the slopes became an important part of the application of this method. In the most highly developed forms even the slopes were artificially constructed. Semiramis' hanging gardens in Babylon are a case in point. Terracing is still a generally applied method on steep slopes in the arid parts of the Mediterranean region as a continuation of the ancient traditions. This age-old tradition is also alive in the Middle East and is attested to by the remains of structures built more than two thousand years ago in the Negev Desert. The technique is used today in a more modern form in agricultural areas on the plains lying along the eastern shore of the Red Sea, for example, at Tihama.

Rainwater harvesting for use in farming flourished in the Negev desert during the days of the Roman Empire, when the Roman trade route to the East passed through the area. Conditions in the Negev were favourable to rainwater harvesting for farming because there was a reasonable rainfall with mean annual values about 200 mm in the part of the Negev considered here. There was a moisture-retentive loamy soil known as loess; and catchments which were large compared to the cultivated area. Evenari and others (1971) have reconstructed some of these ancient farms and proved the possibility of

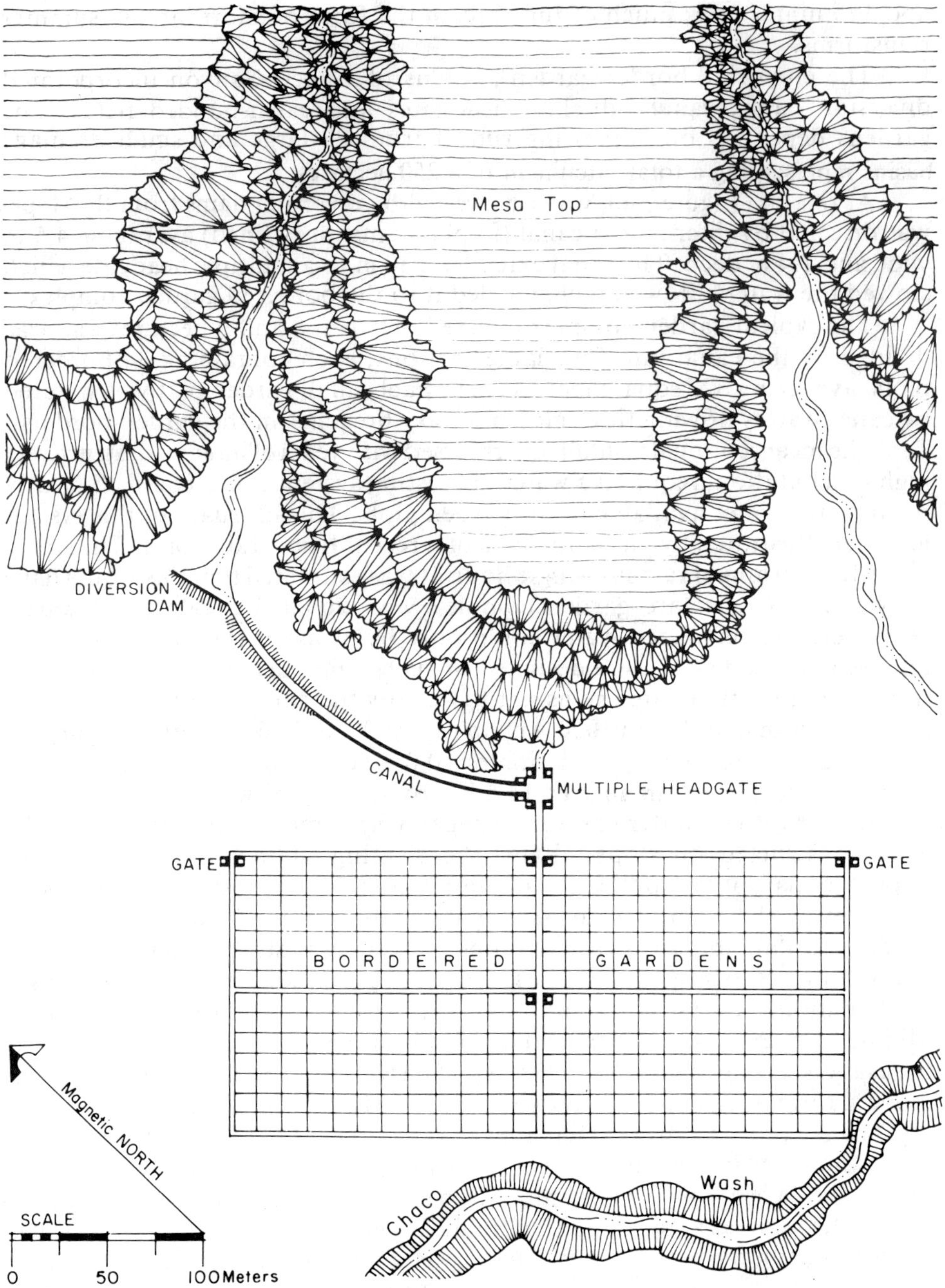

Figure 2-6 Relic water control system in Chaco Canyon with a diversion dam, canal, gate complex and bordered gardens (Vivian, 1974)

successful cultivation. Water for the farms was collected by long canals cut on hillslopes, which intercepted excess rainwater as runoff. In addition, the runoff from the catchment of the farm itself was used. The canals functioned more or less like the channels seen in India today; each had an average cross-sectional area of 0.1 m^2. The biggest canal had a cross-section of 0.4m^2.

The catchment area of an interceptor canal averaged 0.1 to 0.3 ha. The bigger canals drained an area of 1 to 1.5 ha above them. The biggest farm had an area of 5 ha, and received rainwater from a catchment 17 to 30 times its own area. In the case of a very large catchment, a diversion weir was built across the stream which drained it, and the water was then diverted to the farm by another canal. It is estimated that about 2 per cent of the rainfall over a large catchment was utilized, whereas the figure was 10 per cent for small catchments. The Bedouins in the Negev still use these water harvesting techniques. Figure 2-6 shows a typical ancient rainwater harvesting system for farming.

The silt trap is another very useful water harvesting technique which was practised in ancient agricultural communities. These simple stone dams were used to trap silt and water. After seepage of the water into the earth, the moist silted layer was then farmed. The remains of ancient silt traps have been recorded throughout the American southwest, as well as at numerous sites in

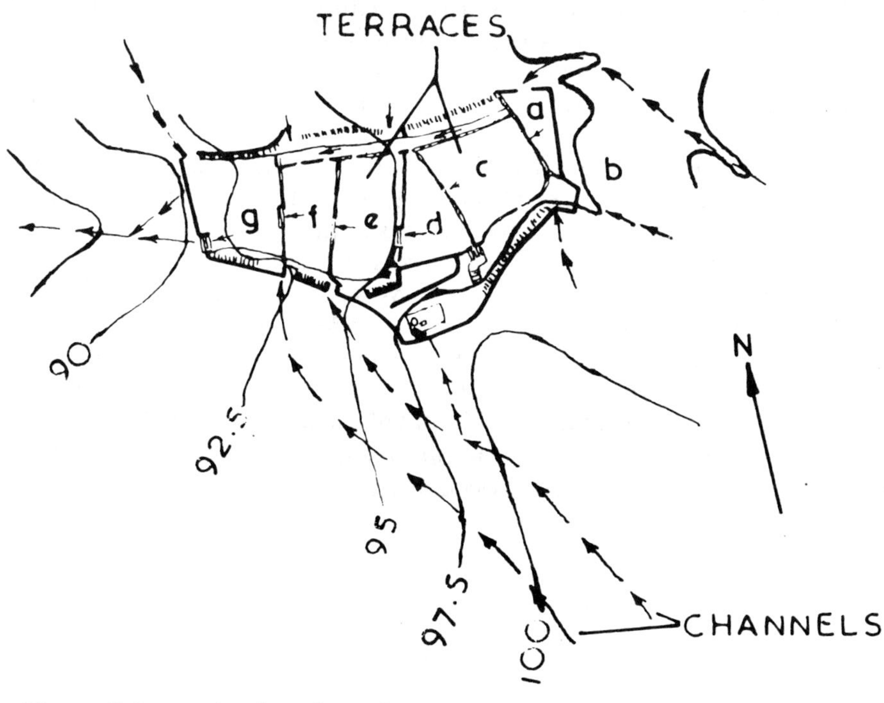

Figure 2-7 Ancient farm in the Negev Desert (Evenari *et al,* 1971)

northern and central Mexico, including Mesa Verde (Colorado), Point of Pines (Arizona) in the American southwest, and at the Rio Gavilan (Chihuahua), the Tehuacan Valley (Puebla) and the Nochixtlan Valley (Oaxaca) in Mexico.

At Chapin Mesa in the Mesa Verde area (Colorado) over 900 silt traps were recorded in a field survey during the 1950s (Rohn, 1963). Archaeologists found 39 series of silt traps, controlling all or most of the intermittent water flow of the area's drainage ways.

One example, Site 800, contains 43 check-dams located in a V-shaped 20 per cent grade canyon. Materials used in the small stone and earth structures were taken from the canyon's sandstone walls. It is possible that brush dams were also used where relief was less steep; but none of these survive. The bases of the Chapin Mesa dam walls rest on bedrock; although they are somewhat short and low, averaging about five to six metres long and 35 to 45 cm high, it is possible that while in use they may have been even higher.

Accounting for possible deterioration or destruction of the structures since the 14th century, it is hypothesized that there may have been as many as 55 silt traps in this Canyon, similar in size, and spaced approximately four to six metres apart. These 55 silt traps created a total of approximately 0.14 hectare of cultivable land; probably enough to support one household. Complementing the silt trap acreage were contour terraces stretching out around the hillside from the wash. In total, including the contour terrace, 250 metres of stone walls were constructed in this Canyon.

Evidence attesting to the capacity of these structures to retain alluvium and runoff can be found in the present-day stands of grass and brush still to be found behind the dams.

A conservative estimate of the total cultivable land created by the 900 silt traps at Chapin Mesa is a figure of between 9 and 14 hectares of top quality soil with a high moisture retaining capacity. It is important to note that silt traps apparently reached their peak between 1150 and 1300 AD when the population of both Chapin Mesa and the Mesa Verde region as a whole had reached its zenith. As with contour terracing in Mexico, it is possible that this water harvesting strategy was a response to increasing population pressures on the existing farmlands of the mesa top.

Finally, it should be noted that Chapin Mesa, as well as the Mesa Verde as a whole, were abandoned by their inhabitants in the 14th century after almost one thousand years of continuous occupation.

CHAPTER 3

Collection and Storage of Rainwater from Roofs

INTERCEPTION of rainwater before it reaches the ground has the advantage that the water may be collected without many contaminants and that the water generally will be suitable for meeting most domestic requirements. The quantity of water that can be harvested is determined by the amount of rainfall, and the size of the area from which the water is collected. For example, 10 mm of rainfall can provide 100,000 litres of water per hectare.

Some methods of water interception in general use include the use of roofs, courtyards and ground (surface) catchments. There is an extensive modern literature on ground catchments (see chapter 4); here we are particularly concerned with roofs.

The use of roofs for water collection is widely practised throughout the tropical world. Corrugated galvanized iron roofs have been used to harvest rainwater in many humid and sub-humid regions. These roofs are cheap and, when used with gutters, can collect water without excessive maintenance costs (figure 3-1). Such roofs are also durable, except in coastal regions where they tend to rust because of salt action.

Costs may be kept down by the increased use of local materials, such as tiles which can be produced on a self-help basis. Tile roofs are very durable and require little maintenance (figure 3-2). Tiles also make less noise during rains than iron sheets, but their weight requires a stronger supporting frame.

In rural areas of many developing countries, thatched roofs are common and can be used to intercept water in small quantities when gutters are installed. However, the water collected is coloured and unattractive, and it is easily contaminated. Thatched roofs are less durable than corrugated iron sheets or tiles. They can, however, be improved by using plastic sheeting which provides a good catchment surface for water harvesting. But, although cheap in cost, such sheeting is not durable. It is easily torn and may be covered by algal growth and probably requires replacement at short intervals. Roof catchments made from cement, bituminous paper and sisal-reinforced materials are in use in some countries. Under satisfactory climatic conditions, collection of rainwater from roofs, when efficiently done, can supplement other water supplies effectively. For example, where the average rainfall is 1800 mm (with an intensity of 0.2 mm per minute as in the Gusii Highlands of Kenya), using 5000 litre storage tanks, enough water can be collected in 12 hours to serve a family of 6 for 45 days. Where there are many rainy days per year (150 in this example) large quantities of water can be collected.

Rainwater collection from rooftops is today widely practised in Africa,

Figure 3-1 A permanent house with pitched roof without gable with a gutter for collecting water in Uganda. Note that the head of the water in the tank has been sufficiently raised in order to enable water to flow by gravity from the tank into the house

Israel and in parts of Rajasthan in India, which has one of the smallest per capita surface water flows in the world.

Bermuda, an island with an average annual rainfall of 143 cm, depends almost entirely on precipitation harvesting schemes for its water. Government regulations ensure that all buildings have properly constructed roofs, gutters and storage tanks. Each house uses water collected from its own roof which is stored in a cistern beneath the house. The systems are said to be able to supply a per capita demand of 80 litres daily. These private schemes are augmented by government schemes with concrete catchments, from which water is trucked in dry spells. But occasionally the water supply system does require augmentation from outside the island.

SEPARATION

The first rainwater running off roofs or courtyards normally contains a lot of sediment. For health reasons and also to keep the water potable, this water should be separated from the supply to be stored.

The simplest device for separating the first flush of rainwater may be a simple throw-over valve, or even a plastic hose which can be deflected from the storage tank. However, such systems are quite unreliable. Simple automatic systems have been devised and they are easy to install. An interesting example from Australia (figure 3-3) is particularly well-suited for roof catchments. This device is a swing funnel made out of sheet metal. The first flush of rainwater falls into "A" causing an increase in weight because the exit drain is very slow. The funnel falls towards the wall and allows clean water to flow through "B" into the storage tank. After some time the slow leak from "A" allows the funnel to fall back into the ready position in time for the next rainstorm.

Another device is a simple baffle tank (figure 3-4) where the runoff passes through a baffle and a sieve before going over a standpipe. Sediments settle in the bottom of the box which needs periodic cleaning.

Figure 3-2 Brick house in Uganda with a tile roof without gable used for collecting water from the roof into a large uncovered brick tank

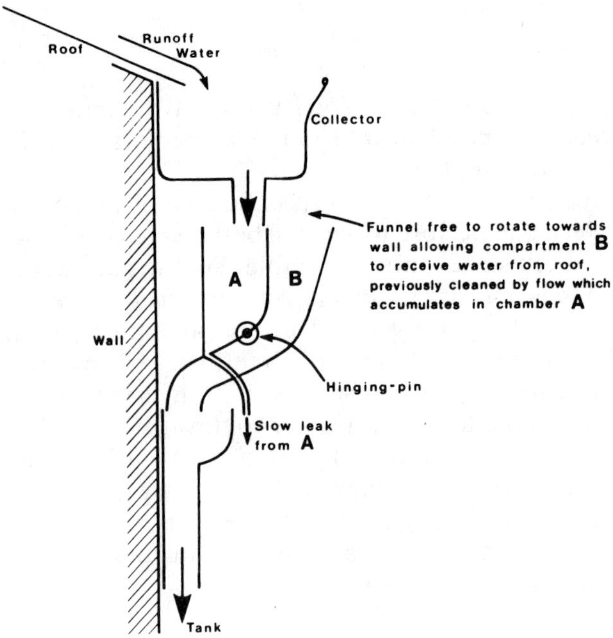

Figure 3-3 Swing funnel for separation of first rainwater from the later cleaner runoff

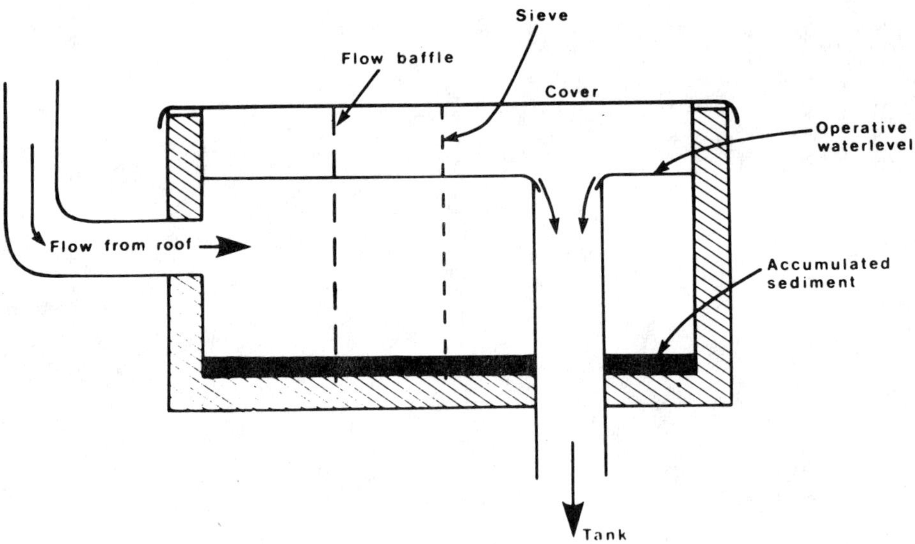

Figure 3-4 Baffle tank to allow grit and dirt to settle from the first flush of water off a roof

FILTRATION AND STORAGE

Water collected especially from ground catchments can be cleaned by using silt traps and sand filters prior to storage. A number of simple sieves have been designed that separate debris from harvested rainwater (figure 3-5). For separation of fine particles, simple filters are used which generally consist of a container filled with layers of sand and gravel; good examples can be seen at the UNICEF Village Technology Unit in Karen, Kenya. Such filters must be cleaned out or renewed occasionally. More permanent filters can be constructed out of cement and incorporated into house construction, in line with storage tanks or cisterns.

Figure 3-5 A galvanized wire mesh acts as a sieve for tanks in part of Gusii Highlands, Kenya

Storage facilities can be above or below the ground. Whichever type is used, it is recommended that the container selected be fully covered to prevent contamination arising from dust, humans and animals as well as to prevent algal growth and the breeding of mosquito larvae (figure 3-6). Open containers (figure 3-7) or household ponds are not recommended for reasons of health and because of high loss due to evaporation.

In the Negev Desert, rainwater from rooftops, courtyards, and streets (Evenari *et al.* 1971) is collected in settling tanks where the silt is deposited. The clear water is then channelled into underground cisterns in the house. These cisterns are normally bottle-shaped (figure 3-8). The capacity of each cistern is in the range of 5 to $10m^3$ (against the average water requirement of 1.5 m^3 per person, 1 m^3 per camel, etc.). The cisterns are lined with stone to protect the inner wall. Public places in the Negev like churches (and forts in ancient days) had open cisterns of larger capacity and they were supplied by canals 1 to 4 km in length often cut directly into hillslopes. These cisterns had stone walls about 1 m high at their periphery to keep animals out, and were provided with settling tanks to remove silt.

Underground storage facilities have the advantage of generally being cool. There can also be a saving in space and cost of construction because the container can be moulded directly in the ground by simply compacting the

Figure 3-6 A brick tank well-sealed against dust or pollutants in Nyeri, Kenya

Figure 3-7 This brick tank is not sealed. Dust and other impurities could easily get into the stored water at this primary school in Gusii District, Kenya

earth, using cement applied by hand, or by draping plastic sheeting on the walls of the excavation. Although cement is the ideal material to use, in some locations natural rock may be excavated to provide underground cisterns.

A good example of a moulded underground container where the soil is just compacted is shown in figure 3-9. This is an underground storage well common in those parts of China where the soil is suitable for moulding. During the earliest days of the nineteenth century, rainwater intercepted by roofs was stored in barrels or corrugated iron tanks. In many cities, particularly those where the reticulated water is not particularly pure, rainwater tanks are still a familiar appendage to houses to ensure a more palatable supply for drinking and cooking as well as a more satisfactory medium for washing. Figures 3-10, 3-11 and 3-12 show the usual arrangements for such corrugated iron tanks which are directly fed by overhead spouting from a nearby roof and placed in the shade of trees in order to keep the water cool.

Larger storage receptacles historically were also fashioned from stone and brick cemented with a suitable waterproof mortar. These were in many cases also installed underground, mainly to minimize heating of the water in summer. Today, underground tanks made of reinforced concrete are easily installed using a double-walled steel form. This circular form is

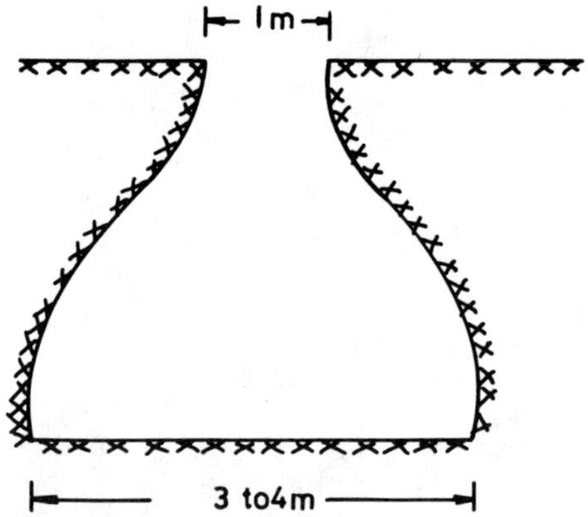

Figure 3-8 Underground cistern for storing drinking water in the Negev Desert

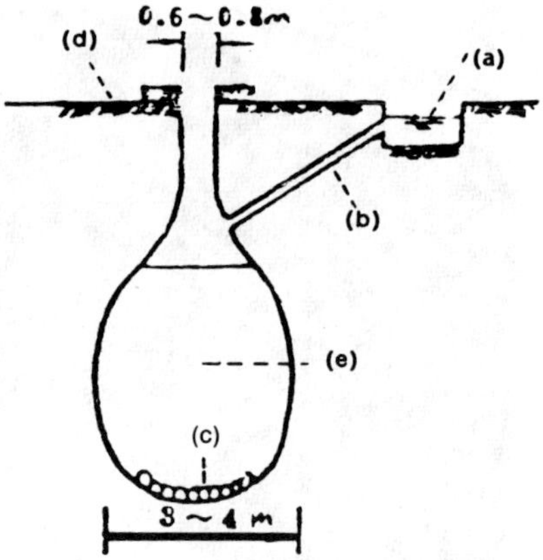

Figure 3-9 Underground well for storing rainwater collected from courtyard areas in China. (a) silt basin; (b) inflow water pipe; (c) protection bed layer; (d) platform at mouth of well; (e) stored water

Figure 3-10 House with corrugated iron pitched roof with a ribbed galvanized iron sheet tank for collecting water in Lesotho

Figure 3-11 Two galvanized iron tanks used for collecting water from two different houses in parts of Zambia. This enables a greater quantity of water to be stored during heavy rains

lowered into an excavation and concrete is poured into the space between the walls. The inner and outer steel wall is removed a day later leaving a large round concrete tank. Reinforcing is usually fixed prior to pouring the concrete; the bottom floor and top cover are added and the excavation is filled. Such tanks represent a substantial initial investment, but the final product has the advantage of extreme durability and potentially large size. In Australia, the most common size has a volume of 100,000 litres. Because of their strength, they can be used as part of a structural foundation. When set in-line with sand filters and settling tanks, and equipped with hand or power pumps they can be used to create completely independent, long-term water storage facilities.

Here there is a wide choice between materials depending on the range of socio-economic conditions as well as environmental requirements. The most common material used is galvanized sheet iron which can be readily riveted and soft-soldered. However, unless these tanks are structurally reinforced by either a steel or wooden framework, they can suffer considerable deformation when filled, and this limits their maximum capacity to about 1,000 litres. Tanks of this type have been widely used since the nineteenth century. Such tanks could probably be incorporated as part of the wall structure of many types of houses but periodic inspection for rust is necessary, and maintenance must be carried out promptly.

Figure 3-12 Corrugated, galvanized iron tanks shaded by trees in Australia in order to keep stored water cool. The lifetime of such tanks, although limited by corrosion, can be greatly extended by lining with cement

This type of tank is a very common type of above-ground storage container. Tanks with capacities of up to 10,000 litres are found in Africa and many other regions.

Depending on availability, either stone or brick can be used to build walls bonded either by lime-mortar or cement mix. Where large tanks are required, certainly when their height exceeds two metres, peripheral reinforcing is simply done by tightening a steel band around the outer circumference. The roofing of this type of tank should be of impervious sheeting such as galvanized iron supported by a wooden framework, as shown in figure 3-13.

Figure 3-13 Concrete and brick walled tank, shaded by trees and roofed with galvanized iron sheets supported by concealed wooden beams

In areas where clay is available, as in many parts of Africa, Asia and Latin America, fired clay pots provide an elegant solution to water storage problems. The main disadvantage would appear to be the limited volume and permeability of the container. Possible developments might include simplified glazing techniques to increase the impermeability of the pots. However, because of their porosity the cooling effect is an advantage that should not be overlooked.

There is considerable scope for designing novel clay storage units which may be incorporated into the wall design of domestic buildings. Such water storage containers have been designed by the UNICEF Village Technology Unit. This is the Ghala Tank which is made by moulding cement inside and outside over a large granary basket (figure 3-14). If the cement wall thickness is taken up to 3 cm, large containers up to 2500 litres capacity can be built. A simple delivery pipe is positioned inside the concrete base during construction.

Wooden barrels are also of use in rainwater storage. They were common in Europe, North America and Australia many years ago. Their disadvantages (limited volume and susceptibility to rot) far outweigh their advantages. Consequently they are now seldom recommended.

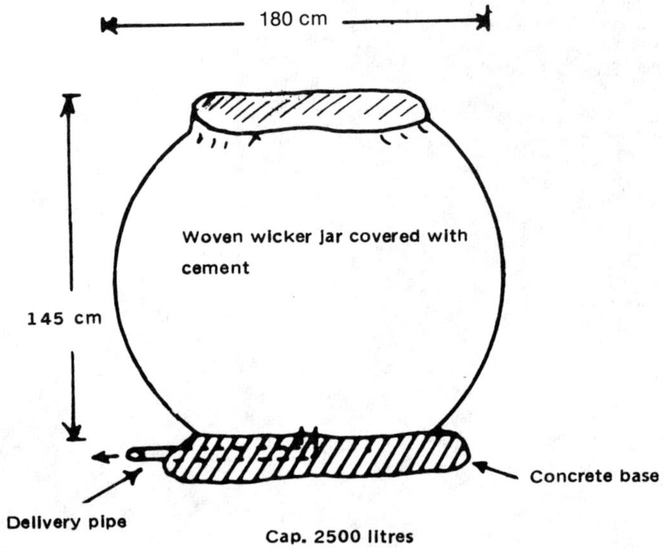

Figure 3-14 Ghala Tank made by moulding cement inside and outside a large woven granary basket

WATER TREATMENT

Water harvested through adequate catchment provisions and stored under safe conditions often does not require any treatment before use in the domestic water supply. A light treatment using a simple device such as pot chlorination may be provided as a safety precaution. When new storage containers made from cement are used for rainwater storage, they can be "cured" by washing with a weak acid or vinegar.

The rapid filtration of rainwater before it is stored is effective for removal of grit and debris, but it seldom removes bacteria if the water is heavily contaminated, and it is always desirable to boil this water before drinking. This practice may be replaced in the future by distilling water using simple solar stills, but at present such a technique is not readily available. For small quantities of water, pot filtration in household trickle filters prior to use gives satisfactory results. Such filters (figure 3-15) can be easily maintained. Larger quantities of water such as that stored in underground tanks can be effectively treated in long-term storage by slow sand filtration.

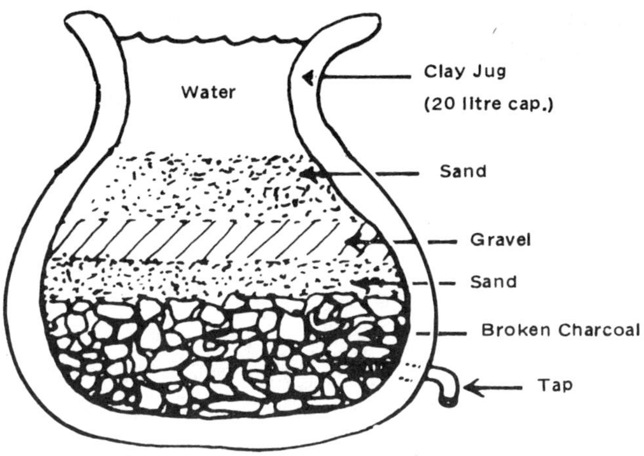

Figure 3-15 Pot filter shown in cross section

Treatment of small quantities of water can also be done by using indigenous plants and natural products. This has been in traditional use in some countries for many years. Certain soils, such as the clay known as "clarifying earth" in the Gezira and Northern Province in Sudan, produce floc in turbid water and induce sedimentation. A similar flocculation of suspended solids can be achieved by adding certain pounded or crushed plants to the water. This is a common practice in Sudan and the coastal regions in East Africa (Jahn, 1976).

DISTRIBUTION

Sometimes rainwater is stored in catchment tanks or large basins in uphill areas where it is gravity-fed to the point of use. In these cases the bacteriological safety of the water must receive particular attention.

More often, the water has to be lifted from the storage facility. In this regard, rope and bucket, endless chains with discs rotated over a winch axis, and similar devices have all been used and, in most cases, contamination of the water is likely, and alternative methods should be used. Hand pumps are particularly useful. They can be permanently installed and have a long service life if they have low use conditions.

The use of harvested rainwater by ordinary rural households would not require any elaborate distribution scheme. But in community-operated catchment projects for domestic water supply, the water can be taken out through a pipe embedded in the embankment, subjected to suitable treatment (e.g. slow sand filtration) and conveyed to the village in a pipe or covered channel. The system can work by gravity, if the embankment is built on ground higher than the village. Alternatively, if individual houses have their own underground storage tanks built to store rainwater harvested from rooftops, they can simply be replenished seasonally.

CHAPTER 4

Collection of Rainwater from Ground Catchments

GROUND CATCHMENTS

RAIN harvesting from ground catchments can be used to supply water for crops, wildlife and livestock as well as for domestic use. The principles involved in constructing such catchments do not greatly depend on the use which is eventually made of the water.

Because of the simplicity of such schemes they have been used in one form or another since ancient times. In the Negev Desert some 4,000 years ago, for example, irrigation water was supplied in an area with an average annual rainfall of about 92 mm. Hillsides were cleared of rock and gravel to increase runoff and ditches were constructed to carry the water to fields below (Evanari *et al.*, 1971).

There are many examples available of various techniques which have been used in the construction of rain harvesting schemes (Johnson, 1979). A fair number of harvesting schemes using ground catchments have been described and a summary of the experience gained in previous projects is useful. It is worth emphasizing, however, that advances in technology, particularly in the production of low-cost waterproofing materials in some cases, make the experience of past construction methods of limited relevance.

The various means of increasing the runoff from an area can be classified as follows:

(a) clearing sloping surfaces of vegetation and loose material;
(b) improving vegetation management by changing ground cover;
(c) mechanical treatment, including smoothing and compacting the surface, contour terracing and microwatersheds;
(d) reducing soil permeability by the application of chemicals;
(e) surface-binding treatments to permeate and seal the surface;
(f) covering the catchment with a rigid surface; and
(g) covering the catchment with a flexible surface.

Surface Clearing

The simplest means of improving the runoff from an area is simply to clear the surface. In ideal situations, very little effort is required. Where the area is fairly impermeable, such as a rock catchment, virtually all the runoff can be utilized if the material which interferes with the runoff is removed.

Removing obstructions can be expected to increase the velocity of the runoff. Unless the natural surface is quite hard, erosion can be expected to increase after the surface is cleared. This simple method may not be appropriate for domestic water supplies if the effect of such erosion will greatly affect water quality. After clearing, the surface can also be contoured and compacted to achieve a greater efficiency in runoff collection. The V-shaped contour producing a 'roaded' catchment is common in Australia and the USA.

Vegetation Management

Runoff from grass-covered areas tends to be greater than that from forest and brush covered lands (Meyers, 1969), and it appears that this approach can increase runoff without appreciably increasing erosion. Research in Utah and Colorado indicated that runoff could be increased from 10 to 20 cm by converting the vegetation from trees (aspen) to grass (Beattie, 1969).

Vegetation management is effective in increasing runoff in areas receiving 280 mm or more of annual rainfall. Furthermore, the conversion efficiency for increased runoff improves with increased rainfall up to 860 mm (Cooley *et al.*, 1975).

Aside from rainfall, harvestable water yields depend on several factors including the percentage of total precipitation occurring as snow, the type and depth of the soils, slope gradients, and the varieties of plants with their associated evapotranspiration rates.

According to experiments carried out since the 1950s in Arizona, USA, possible vegetation management strategies include: conversion of areas immediately adjacent to stream channels to runoff-enhancing vegetation covers; clearing the forest or shrub cover in uniform or irregular strip cuts; and thinning overstory densities.

On the basis of studies done on experimental watersheds in areas of mixed conifer, ponderosa pine and chaparral, researchers estimate that if a vegetation management programme were implemented in Arizona's 15 major drainage regions, total potential water yields would increase (under average rainfall conditions) by approximately 600,000 to 1,200,000 acre-feet a year (Folliott and Thorud, 1975).

However, the feasibility of such a project would be tempered by the competing demands of other uses; by the presence of areas below the conversion threshold; and by numerous physiographic constraints.

In spite of these constraints, the researchers advocate active consideration of vegetation management for large areas of the American southwest. More work is needed, however, on the transport of harvested water from collection points to user areas; on the extrapolation of data from experimental watersheds to other locations; on better site inventories; and finally, on vegetation-managed water harvesting within multi-purpose planning frameworks.

Another informative vegetation management experiment was carried out in an experimental forest in southern California. Conversion of brush to grass cover for the purpose of increasing livestock forage led to severe soil slip erosion, flooding and debris-filled reservoirs in the San Dimas watershed.

Conversion to grass cover on steep slopes adjacent to stream channels is one of the most effective means of increasing stream runoff; yet as demonstrated in the case of the San Dimas watershed, this practice can result in potentially serious environmental consequences (Rice and Foggin, 1971).

Another concern has been raised in the case of the use of herbicides to kill the original vegetation cover. The deleterious effects upon human health by certain herbicides are just coming to light.

One final concern relates to the cost-effectiveness of grass cover in comparison with other harvesting catchment covers. An experiment in Arizona comparing the capacities of three different catchment surfaces—bare compacted earth, wax, and grass—showed that grass was the least effective of the three methods in delivering runoff to forage plots. Its use was discontinued after two years (Frasier and Schreiber, 1978).

One countervailing factor is the greater utility of grass cover over bare compacted earth and wax. It may be that a low-cost low-efficiency cover has greater overall utility than the other two options.

In sum, it can be said that the use of vegetation management for water harvesting contains both promise and problems. Further experimentation is needed before it can be detailed with precision.

Mechanical Treatment

Smoothing and compacting a surface will eliminate losses due to infiltration and depression storage of rainwater and hence will increase runoff. Conventional construction equipment (graders and rollers) can provide such treatment quickly and fairly cheaply. Sealing fissures on a rock surface can also be a simple way of increasing the effectiveness of the rock catchment.

Erosion remains a possible problem if the material of the catchment, even when compacted, is not water resistant.

Current research is being carried out in the southwestern United States where contour terraces (termed variously 'desert contour strips' or 'conservation bench terraces'), as shown in figures 4-1 and 4-2, are constructed along a slope perpendicular to runoff flow. The terraces are separated by sloping collector areas which provide runoff for the narrow field strips below them. The main principle underlying this technique is the use of level, ridged fields to control erosion and to retain, spread and infiltrate storm runoff from the upslope collector areas.

Experimental research and a computer simulation model provide us with a fairly accurate picture of the potential of this system in arid southern Arizona. The proponents of this technique argue that considerable runoff occurs on slopes above stream channels and on-site use of this water eliminates the losses inherent in its collection, storage and distribution. Another clear

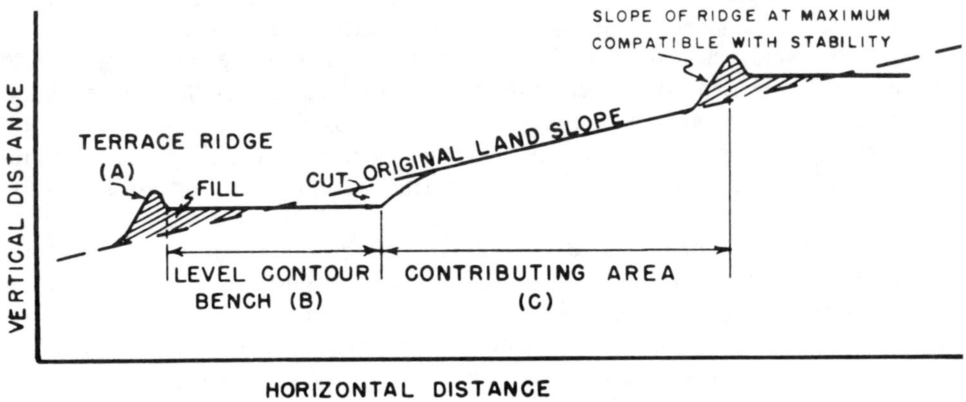

Figure 4-1 Cross section of a conservation bench terrace (Jones and Hauser, 1975)

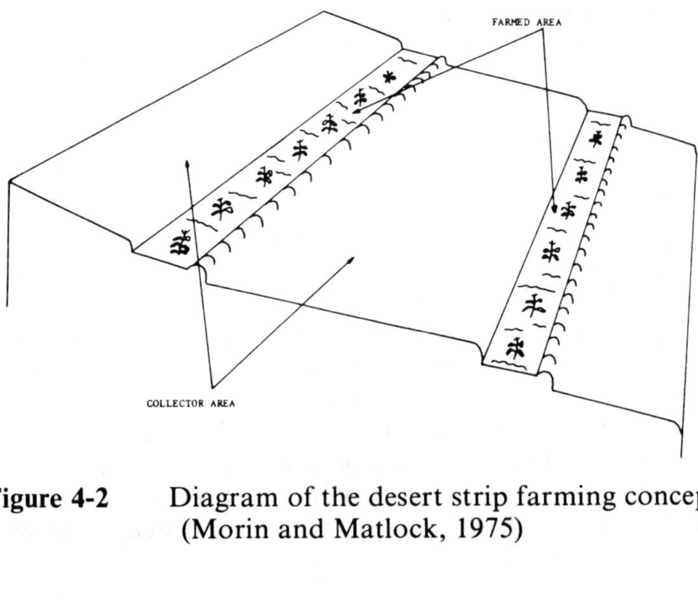

Figure 4-2 Diagram of the desert strip farming concept (Morin and Matlock, 1975)

advantage of this technique is its very modest requirements in terms of labour and capital investment (Morin *et al.,* 1973; Morin and Matlock, 1975).

The catchment area can either be left in its natural state, or cleared and treated to enhance runoff, or planted to range crops. As with other contour systems, provisions are made to distribute runoff evenly on the field strip as well as to allow excess runoff to flow to the lower collector area and field.

Field tests conducted during the early 1970s in the Atterbury Watershed in Arizona, an area which normally receives about 140 mm of rainfall in the summer months, demonstrated that significant harvests of short-season grain sorghum (a crop requiring 570 mm of rainfall to mature) were achieved by means of this contour strip technique.

As shown in table 4-1, investigators found that over a period of three years which experienced widely different amounts and patterns of rainfall, grain sorghum yields ranged from 0 to 4,400 kg per hectare. (Average yields of grain sorghum under irrigation are 4,500 kg per hectare.)

Table 4-1 Actual yields of grain sorghum at Atterbury Watershed during the 1970, 1971, and 1972 growing seasons

Year	Rainfall mm	Yield	
		Range kg/ha	Average kg/ha
1970	190	800-2,300	1,600
1971	246	800-4,400	2,600
1972	137	0-1,100	500

Source: Morin and Matlock, 1975

These field experiments indicate that timing of rainfall and field soil moisture conditions at the time of germination are as important a factor for successful crop production as the total amounts of precipitation.

On the basis of the computer model, investigators were able to simulate long-term productivity of a contour strip system in the Tucson basin. The basic relations of the model are diagrammed in figure 4-3. The model inputs include data on rainfall, runoff, soil moisture conditions, evapotranspiration, temperature, soil and crop characteristics.

The model indicates that, given an optimum 12 : 1 collector-area to farmed-area ratio, significant yields of short-season grain sorghum will be achieved in four out of every five years. Additionally, the model and field study show that crop failures will inevitably occur in bad rainfall years. However, if the collector areas are given over to livestock grazing, then the loss in productivity of the total system is quite limited.

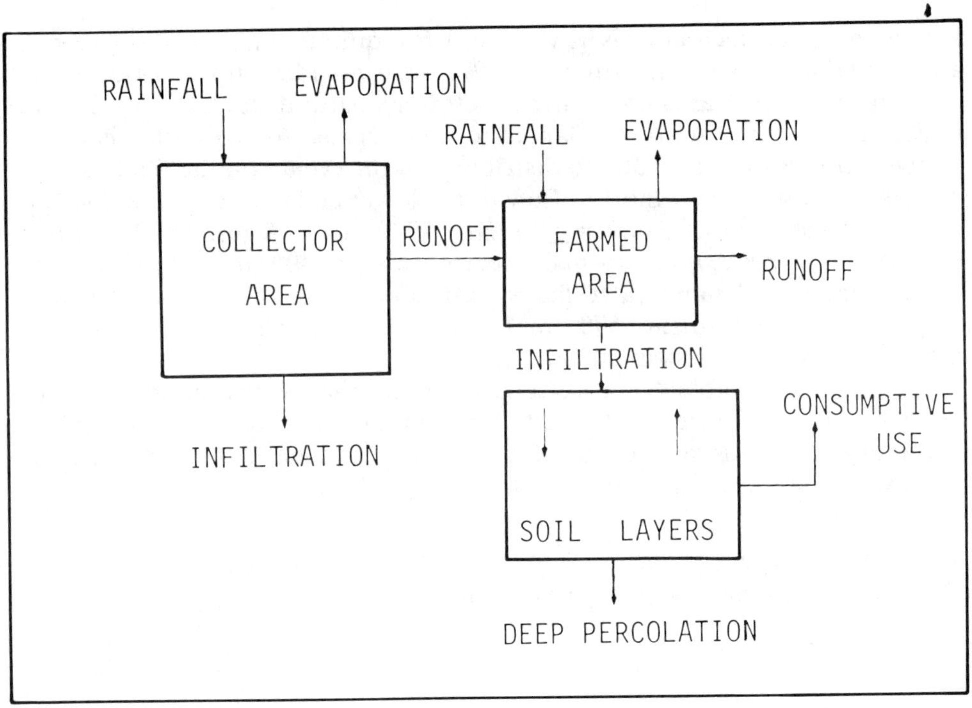

Figure 4-3 Diagram of the desert strip farming concept (Morin and Matlock, 1975)

Another series of experiments carried out over fourteen years at the Southwestern Great Plains Research Centre at Bushland, Texas (USA) clearly demonstrates that modern contour terraces (termed, in this case, conservation bench terraces) can increase available water and crop yields significantly on gentle slopes in dryland regions (Jones and Hauser, 1975).

Average annual precipitation in Bushland is 466 mm and average April to September evaporation from water surfaces is 1,300 mm. Topography is nearly flat and treeless with natural drainage flowing to shallow depressions (*playas*). The predominant soil is Pullman clay loam.

The experimental area contained both conservation bench terraces which were continuously cropped with grain sorghum and graded bench terraces which were cropped in a wheat-sorghum-fallow sequence.

When yields from these two systems were compared to those from sloping plots, it was shown that bench levelling increased mean annual sorghum yields by 43 per cent; and contour bench terraces (which received a mean runoff of 70 mm per year from their collector areas) increased mean annual sorghum yields by 80 per cent.

Investigators concluded that the major advantage of bench levelling over the conservation bench terraces was that higher levels of production were achieved because all available land was cropped. This advantage was offset by the greater probability of lower yields. A major advantage of conservation bench terraces is that only one-third of the area requires levelling.

In the Negev Desert, slopes are levelled over short widths at different elevations, with a low mud or stone wall separating terraces (figure 4-4). Such terracing is found on hillslopes everywhere as well as in smaller wadis in the Negev (Evenari *et al.,* 1971). The spacing between walls (i.e. the width of the terraces) is 12 to 15 m in the Negev farms where the walls are 80 cm high and the difference of level between terraces is 50 to 60 cm. Water flows from one terrace to another negotiating a gentle drop in between. On large slopes, the drop will be higher, and a structure has to be provided to dissipate the energy of falling water and avoid erosion of the field. In the Negev, this structure takes the form of a series of short stone steps.

The technique of terracing wadi beds is practised in South Yemen. Here besides rainwater, dew and mist are also collected during night-time to provide substantial amounts of moisture (Anon, 1978). The nearness of South Yemen to the sea as well as to the desert (Rub al Khali) promotes substantial dew formation, as in Israel. The hillslopes of North Yemen where the famed Mocca coffee is grown are cultivated by terracing. Water for these fields is harvested from the runoff on neighbouring uncultivated land. This is said to be so thoroughly done that no rain falling on the hillslopes escapes.

Microwatersheds operate on the same basic principle as other forms of land alteration where runoff from a collector area is concentrated, retained and infiltrated within a small ridged plot. In the case of microwatersheds, the collector area and infiltration plot serve only one individual tree or a very limited number of plants. Microwatersheds constructed for individual trees are often found in combination with other water harvesting structures.

Like other water harvesting systems, the collector area of microwatersheds is devised to maximize runoff while infiltration is encouraged in the basin (termed 'runon' area) immediately surrounding the plant. Frequently, mulch is used to decrease evaporation.

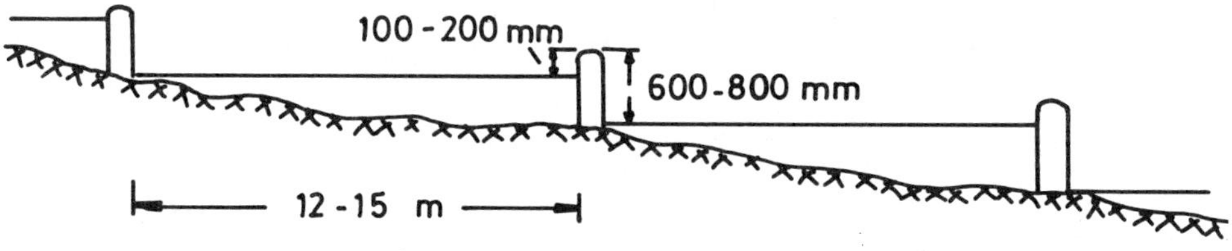

Figure 4-4 Terraces in wadis in the Negev Desert

Experiments at the Central Great Plains Field Station in Akron, Colorado (USA), have demonstrated that minimum runon areas containing a vertical mulched slot deepen penetration of water and reduce evaporation loss possibly by as much as 50 per cent as in figure 4-5, A and B.

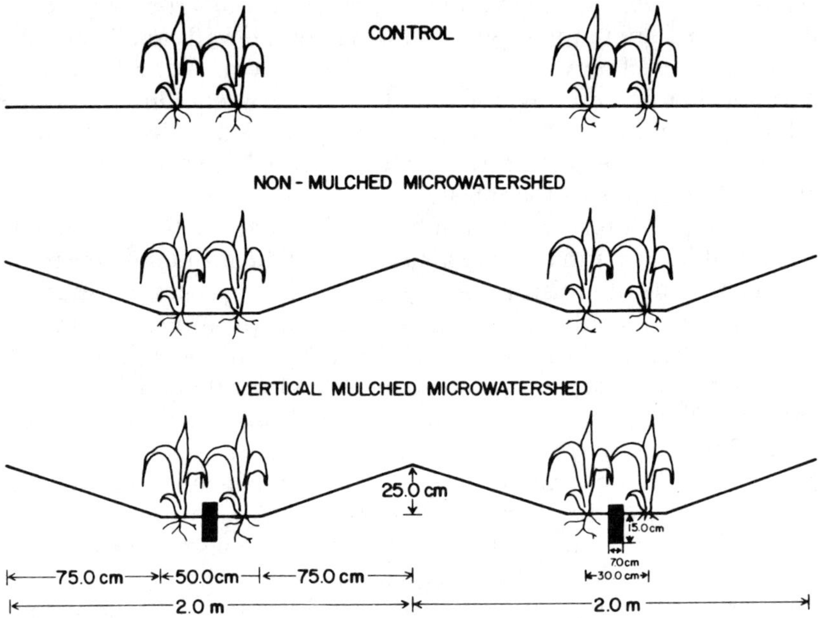

Figure 4-5 Vertical mulched microwatersheds showing (**above**) design and (**below**) spacing and wetting fronts (Fairbourn, 1975; Gardner, 1975)

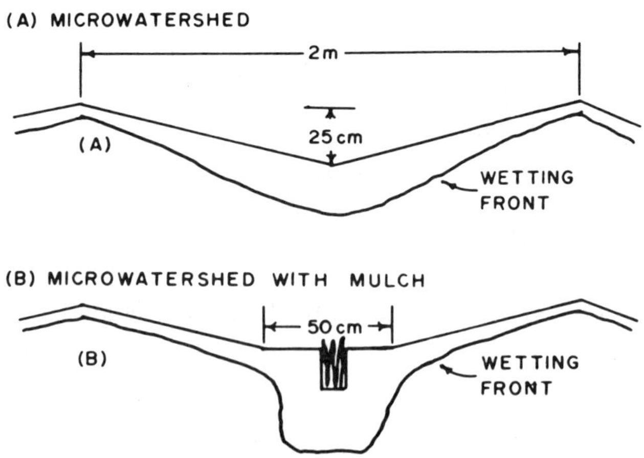

CHEMICAL APPLICATION TO REDUCE SOIL PERMEABILITY

Soil permeability can be reduced by colloidal dispersion and/or hydrophobic treatments (Hillel, 1967). In the former, self-crusting of soils containing clay is caused by the addition of sodium salts which disperse the colloids. The treatment is cheap but recent research has found that erosion of the soil is a severe problem with such treatment and that the salt tends to be washed away in a short period (Myers, 1967). The salt washed away by the runoff would obviously have an adverse effect on the water quality.

Treating minimally vegetated desert soils with sodium can reduce infiltration rates temporarily. Clearing, shaping and compacting the soil prior to or during the sodium applications can result in long-term effectiveness. Compaction of even low-clay, sodium-treated soils can result in a significant increase in available runoff. Additionally, salt is an herbicide (Myers, 1967).

An experiment conducted on Whitehouse loam soils near Tucson, Arizona (USA), in an area receiving 200-400 mm annual precipitation resulted in a 50 per cent runoff over a three year period (Dutt and McCreary, 1975).

Two 1.2 m wide waterways with a two per cent grade were constructed between sodium-treated catchment areas. These areas were cleared and smoothed, and 11,000 kg per hectare of granulated salt was mixed into the upper 5 cm of soil. The soil was then compacted following two light rains (figure 4-6).

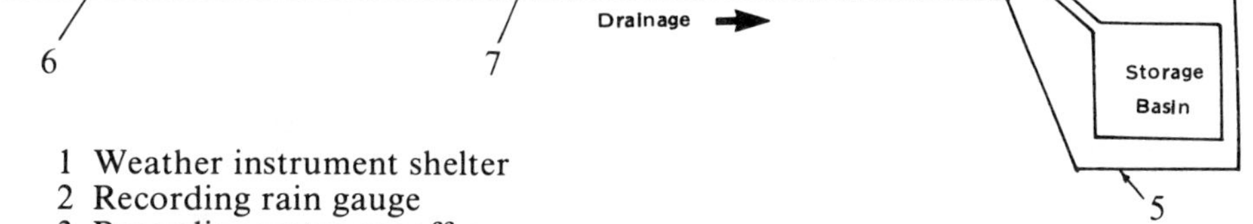

1 Weather instrument shelter
2 Recording rain gauge
3 Recording water runoff gauge
4 Recording tank depth gauge
5 Perimeter fence, 8′ (6″ hogwire, 2′ 2-strand barbed wire)
6 Salt treated area, 39,000 sq. ft.
7 Grapes spaced 8″ apart in untreated 4″ x 4′ trough

Figure 4-6 Page-Trowbridge experimental range salt-treated water harvest system (Dutt and McCreary, 1975)

The waterways were cropped and, additionally, conducted water to a sodium-treated storage tank with a capacity of 340,000 litres. Wine grapes and 57 deciduous fruit trees were planted. Over a three-year period the vines are reported to have done fairly well; no information was given about the trees.

Water quality of the treated tank supply was good, suggesting that after initial establishment, the salt remains in the catchment area. Moreover, the tank did not empty once during the three years of the experiment in spite of two periods of drought.

Salt treatment of soils is appealing due to its low cost (table 4-2). However, its effectiveness can deteriorate after one year unless compaction and shaping are performed. One additional problem with this method is that increased runoff may encourage erosion (Cooley *et al.,* 1975).

Hydrophobic soils are created by the addition of water-repellent materials to reduce the wettability of the soil surface. Many materials can be used, of which the most successful has been silicone. In research in Arizona, Myers *et al.* (1967) found sodium methyl silanclate penetrated the soil to form an inert, hydrophobic resin which is not biodegradable. Silicone treatment experiments involve spraying test sites with an aqueous solution of a silicone-water repellent which reacts with the calcium or magnesium in the soil to form an inert water-repellent resin.

A treated 200 m^2 plot on smoothed sandy loam soil in Arizona initially yielded 94 per cent runoff compared to 33 per cent for an untreated smoothed plot. Repellency dropped to 40 per cent over the next four years, probably due to erosion and weathering; but was restored to 85 per cent by retreatment (Myers and Frasier, 1969).

Subsequent experiments suggest that the combined application of silicone plus a soil stabilizer prolongs high runoff efficiency.

Silicones arc easy to apply and relatively inexpensive (table 4-2). However, the treatment does not work well on soils in which swelling clays are present; it is most appropriate for sandy soils with minimal structural development. One problem with silicone is that it provides no stability and increased runoff can lead to erosion problems (Myers, 1967).

SURFACE BINDING TREATMENTS

Petroleum products which penetrate the surface, bind soil particles together and provide an impermeable surface, have been used in many situations. Extensive research and investigation has been undertaken in Israel, and a limited number of experiments have been carried out in Mexico and the

Table 4-2 Water costs for various water-harvesting treatments

Treatment	Runoff (%)	Estimated life of treatment (years)	Initial treatment cost ($/yd²)	Annual amortized cost[a] ($/yd²)	Water cost in a 20-inch rainfall zone ($/1,000 gall)
Rock outcropping	20-40	20 - 30	< 0.01	< 0.02	0.22 - 0.45
Land clearing	20-30	5 - 10	0.01 - 0.02	< 0.01	0.30 - 0.45
Soil smoothing	25-35	5 - 10	0.05 - 0.07	0.01 - 0.02	0.25 - 0.71
Sodium dispersant	40-70	3 - 5	0.07 - 0.12	0.01 - 0.02	0.13 - 0.45
Silicone water repellents	50-80	3 - 5	0.12 - 0.18	0.02 - 0.04	0.22 - 0.71
Paraffin wax	60-90	5 - 8	0.30 - 0.40	0.05 - 0.10	0.50 - 1.49
Concrete	60-80	20	2.00 - 5.00	0.17 - 0.44	1.89 - 6.53
Gravel covered membranes	70-80	10 - 20	0.50 - 0.70	0.04 - 0.10	0.45 - 1.27
Asphalt fibreglass	85-95	5 - 10	1.00 - 2.00	0.14 - 0.48	1.31 - 5.00
Artificial rubber	90-100	10 - 15	2.00 - 3.00	0.21 - 0.41	1.87 - 4.00
Sheet metal	90-100	20	2.00 - 3.00	0.17 - 0.26	1.51 - 2.57

[a]Based on the life of the treatment at 6 per cent interest.

Source: Frasier, 1975

United States. These experiments indicate that problems arise regarding the durability and cost of fuel oil treatments.

Recent work with Texas crude oil demonstrated that repellency disappeared within six months of spray application. Researchers therefore suggest caution in the use of fuel oils as repellents, particularly in view of their rising cost (Fink and Frasier, 1975).

A single-season comparison of five soil treatments for water harvest radish cultivation in Mexico indicated that excellent crop growth was achieved by means of all five soil treatments. Table 4-3 summarizes the results obtained from treatments including: polyethylene cover, compacted earth, and two types of diesel fuel:

Table 4-3 Radish yields with five soil surface treatments

Soil surface treatments	Yields of radish for the following percentages of area used to collect rainwater			
	25 tons/ha	50 tons/ha	75 tons/ha	Average tons/ha
Polyethylene cover	55.65	63.50	76.72	65.29
Straw cover	48.22	59.00	64.36	57.20
Compacted earth (CE)	54.28	60.58	78.80	64.56
CE diesel treated 250 ml/m^2	49.82	61.86	79.04	63.57
CE diesel treated 125 ml/m^2	54.78	61.45	77.64	64.62
Average[a]	52.55	61.28	75.32	

[a]Check yielded 36.66 ton/ha with no soil surface treatment and no area dedicated to harvest rainwater.

Source: Anaya and Tovara, 1975

The Mexican experiment suggests that in the case of similar yields, the cheapest soil treatment should be selected. In this case, compacted earth or straw cover would be cheaper than the other choices.

Researchers are now studying the potential use of cheaper petroleum distillation residues as catchment coatings (Fink and Frasier, 1975).

Paraffin soil treatment is also a good possibility. Here molten paraffin is sprayed on cleared and smoothed soil surfaces, penetrating the soil up to 25 mm and stabilizing soil particles as it penetrates.

In a recent three-year experiment, more than 2,000 kg per hectare of forage were harvested each year in an area near Tombstone, Arizona (USA), receiving less than 130 mm. Using a runoff area twice the size of the forage plots, the resulting forage yields were approximately 16 times greater than those of untreated control plots (Schreiber and Frasier, 1978).

Pariffin can also be applied in the form of granules or flakes and allowed to melt and spread, forming a surface that, in one experiment, yielded a 90 per cent runoff, compared with 30 per cent runoff from untreated plots, and a 100 per cent runoff from a butyl-covered plot (Fink *et al.,* 1973).

In another experiment, two collector areas, one a 0.4 hectare catchment on a clay loam soil with a slope of 5 to 8 per cent and 300 mm annual precipitation, and another, a 0.3 hectare catchment on a sandy clay loam soil with similar slope and 300 to 400 mm annual precipitation, were both sprayed with melted paraffin after having been graded, sterilized and wet-compacted. Both catchments harvested water at a cost that was competitive with that of hauled or piped water. Moreover, it was found that the method worked best on the sandy soil (Cooley *et al.,* 1976).

These and other experiments (table 4-2) suggest that paraffin-treated soils provide high-quality water, are durable, and low in cost for materials and application (particularly if dry granules are hand-applied) relative to other chemical treatment methods.

At present, research is being conducted to find ways to minimize the loss of repellency from the freeze-thaw effects of cold weather, and also to determine which kinds of paraffin are best suited to a variety of soil conditions (Fink and Mitchell, 1975; Cooley *et al.,* 1976).

Asphalt coatings used in rainwater catchments range from simple to sophisticated. Costs for all asphalt techniques are significantly higher than the other methods reviewed so far (table 4-2).

Early experiments during the 1950s and 1960s demonstrated that the most effective methods consisted of a two-layer spraying of catchments. Sites were first cleared, smoothed and sterilized. A cutback asphalt or bitumen in solvent was then sprayed on the soil, penetrating and making a strong porous pavement. This pavement was then topped with a non-penetrating asphalt emulsion to seal pores and protect the base against deterioration by photo-oxidation.

All such pavements remained in good condition after 2 to 4.5 years of cold weather alternating with very hot, sunny weather; and, with minimal maintenance provided 100 per cent runoff. However, in sunny dry areas, asphalt runoff is often discoloured by asphalt oxidation products. This discolouration is not removable by sand and soil filtration. The quality of water harvested in this manner is judged acceptable for livestock (Myers *et al.,* 1967).

From this rather simple technique several very sophisticated combinations of asphalt and other materials have been developed.

Wind damage to thin plastic and metal films (black polyethylene, polyvinyl fluoride, aluminium foil, chlorinated polyvinyl, and butyl) used as catchment covers can be reduced substantially by bonding the films to sprayed asphalt pavements; although subsequent problems remain with film durability and water quality (Myers, 1967).

Another recent development includes placing layers of fibreglass or polypropylene matting on the soil surface and spraying them with asphalt, then sealing them with roofing-grade asphalt emulsion. Usually little surface preparation is required, and almost any soil is adequate. This method results in a very durable, efficient catchment, with the matting providing the reinforcement and the asphalt providing the waterproofing.

Painting the asphalt, another innovation, protects it from sunlight and reduces runoff discoloration. This is a relatively expensive method (table 4-4).

Advantages cited for asphalt-fibreglass include easy installation, simple maintenance, and durability. Research conducted on nine catchments demonstrated that the techniques can provide a dependable livestock water

Table 4-4 Construction costs for 1,100m² asphalt-fibreglass catchment

Item	Cost	
Plot preparation		
Bulldozer, 6 hr at $ 20	$ 120	
Labour, 14 hr at $ 3.50	49	
Supervision, 10 hr at $6	60	$ 229
Soil sterilant		
Monoborchlorate		
150 lb. at 14 cents	$ 21	
Labout, 2 hr at $3.50	7	
Supervision, 1 hr at $6	6	$ 34
Asphalt-fibreglass		
Fibreglass 1-1/2 oz		
1,200 yd² at 40 cents	$ 480	
SS-2 emulsion, 550 gal		
at 30 cents	165	
Brooms, 3 at $ 5	15	
Labour, 20 hr at $3.50	70	
Supervision, 10 hr at $ 6		$ 790
Seal coat		
Roofing emulsion,		
370 gal at 60 cents	$ 222	
Brooms, 3 at $ 5	15	
Labour, 8 hr at $3.50	28	
Supervision, 4 hr at $ 6	24	$ 289
Total		$ 1,342

Source: Myers and Frasier, 1969

supply in dry rangeland areas (Myers, 1967; Frasier, 1975).

Another method developed recently to utilize the relatively low cost and high runoff efficiency of plastic has been to spread the plastic film on the ground and cover it with a layer of gravel. The gravel protects the film against wind and weathering, but it does reduce runoff efficiency by retaining the water that is lost to evaporation. This method is superior to asphalt catchments as it does not produce potentially toxic phenols. A machine has been developed at the University of Arizona (USA) which extracts gravel from the soil, dispenses plastic, and then covers it with the extracted gravel (Cluff *et al.,* 1969).

One further development along these lines has been the APAC (asphalt-plastic-asphalt-chip-coated) method. Soil is sprayed with asphalt, then a layer of plastic is put down which is covered with another layer of asphalt and topped with a layer of gravel chips.

In 1973, two catchments using this technique were installed on the Papago Indian Reservation in Arizona (USA). Installation was accomplished using an asphalt dispensing truck, a dump truck equipped with a chip spreader, and a nine-man crew. Costs are estimated at $4,940 to $7,160 per hectare. Runoff efficiency is 85-90 per cent with an estimated catchment life of ten to fifteen years. With automated plastic unrolling and an experienced crew, costs can be cut substantially.

Researchers suggest that the APAC method be used where less sophisticated methods are somehow impractical or where a very high runoff efficiency is needed to man a dependable water supply (Cluff, 1975; Cooley *et al.,* 1975).

RIGID SURFACE COVERINGS

The most well-known rain harvesting schemes exist on the island of Gibraltar. One catchment, occupying 4 hectares on the east side of the island, consists of corrugated galvanized iron sheets bolted to a timber framework which rests on piles (Sheppard, 1962). Runoff from the catchment is stored in reservoirs excavated in rock. Concrete can also be used on a catchment (figure 4-7). It is prohibitively expensive on large catchments, but it has a long life expectancy (up to 20 years or more).

In Australia, rigid surface rainwater catchments have been built independently to provide water for people and livestock. Over forty years ago Kenyon (1929) prepared a paper on "ironclad catchments" in which he analysed demand patterns and precipitation records. He proposed a scheme with a 2,300 square metre catchment made of flat sheets of galvanized iron on a timber framework to prevent corrosion by contact with the ground. Similar schemes had already been constructed in the State of Victoria, feeding into concrete tanks (Kenyon, 1929).

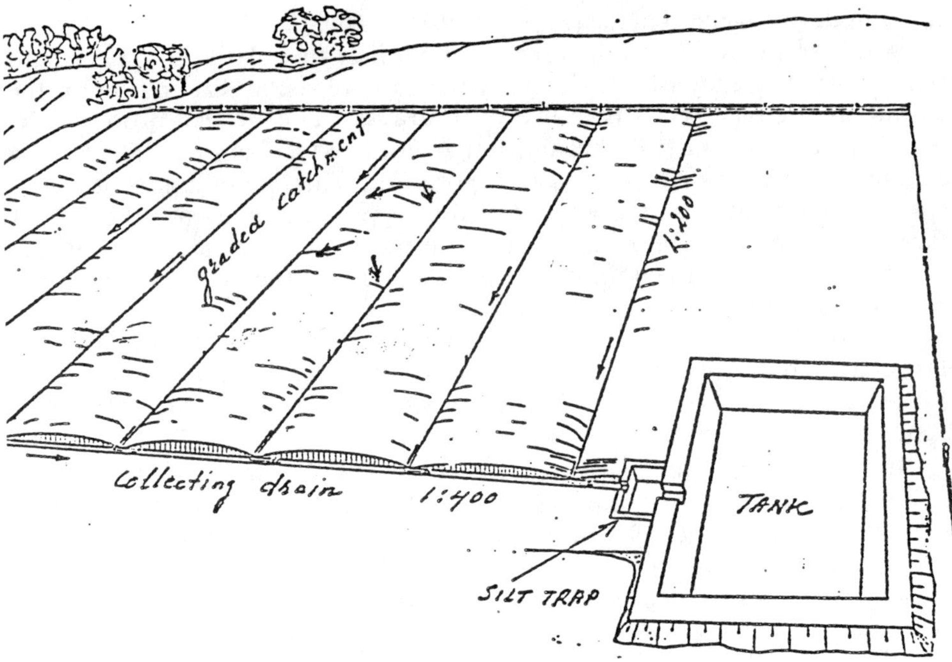

Figure 4-7 Concrete surfaces have an excellent life expectancy as a catch for runoff. Sheet metal also may be used

Along stock routes in South Australia roof catchments of iron or timber substructures were constructed as early as 1885. They continue to be a practical and necessary solution to water supply problems in this area. Three schemes built in 1960 consisted of "rainsheds" of galvanized steel supported on a steel frame about 2 metres above ground giving 650 square metres in area. Two 40,000 litre steel tanks stored water under each shed (Martin, 1910). Sheet metal catchments are also in use in Mexico.

Highway catchments are also being considered as a potential source of harvested runoff for the purposes of livestock water, supplemental irrigation for forage, or highway beautification. At present, much of this water is wasted. However, it is estimated that with the construction of relatively inexpensive diversion ditches and storage structures, significant amounts of runoff can be harvested (Evans *et al.,* 1975).

Preliminary calculations suggest that the interstate highway systems of the State of Wyoming (USA) would provide 2 hectares of catchment per kilometre. With a 90 per cent catchment efficiency and a 250 mm average annual rainfall, the amount of harvested water would be close to 4.7 million litres per km. (Cooley *et al.,* 1975).

Figure 4-8 illustrates the concept of water harvesting from highways. The water captured in this manner can be either diverted to adjacent agricultural fields or it can be used to irrigate the right-of-way which, when properly levelled, fertilized, and seeded can yield up to 2.5 metric tons of hay per highway kilometre in semi-arid Wyoming (USA).

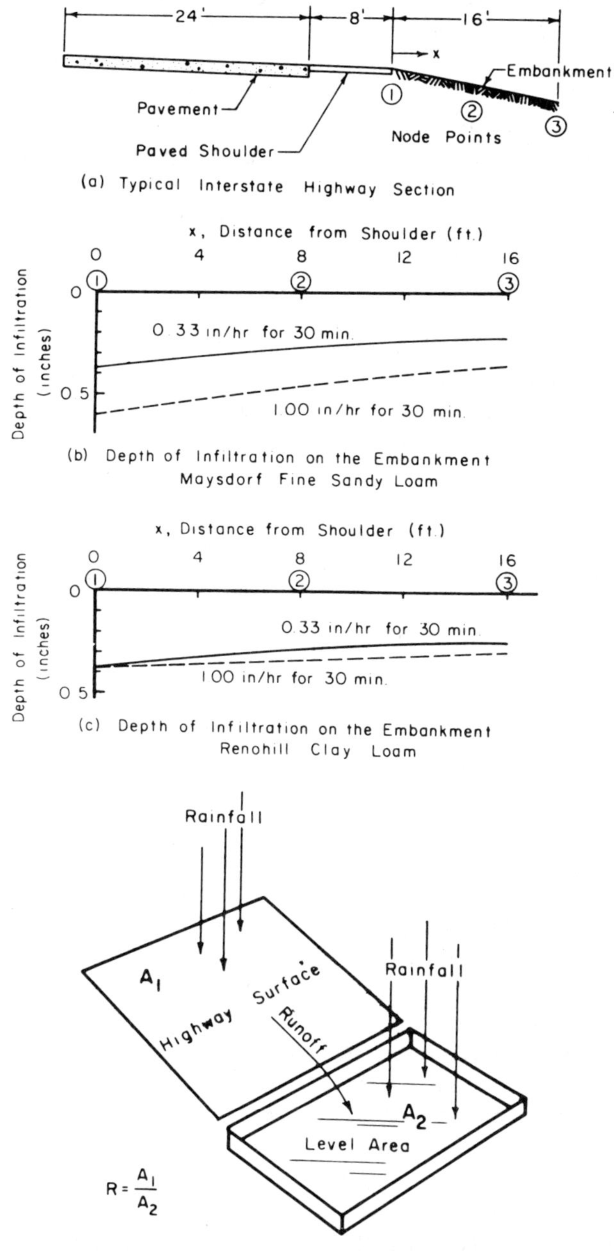

Figure 4-8 Diagram of infiltration on highway embankments and of a highway water harvesting system (Evans *et al,* 1975)

FLEXIBLE SURFACE COVERINGS

During the last decade a variety of prefabricated products have become available which make it possible to quickly and effectively waterproof virtually any area. Earlier coverings were made of materials similar to those used for roofing. One type used in canal linings was a prefabricated fibreglass mat saturated with asphalt and produced in rolls.

As various plastics became available they were tested extensively by the US Bureau of Reclamation for use as canal linings. Results of these tests indicated that no plastic was suitable as an exposed lining, but if covered by at least one foot of material (to prevent exposure of the membrane to air and sunlight) the newer plastics, particularly polyvinyl chloride and polythylene, worked very well in reducing water losses.

Limited experience with expected plastic films as water catchments indicates two principal problems: deterioration of the plastic and wind damage to the lightweight membrane. One way of reducing these problems is to cover the sheets with a layer of material such as gravel. This is an unsatisfactory solution for a catchment surface, however, since possible damage to the sheet cannot be detected and the material used to cover the sheet retards runoff and decreases the effectiveness of the catchment.

Plastic materials therefore appear to be of limited use in waterproofing catchments.

Another interesting possibility is aluminium foil, which was laid in rolls on a hot-sprayed asphalt emulsion in Arizona. One particular catchment gave trouble when individuals walking on the sheet caused pebbles on the unsmoothed base to protrude and rip the aluminium (Myers, 1967). The asphalt bond eliminated problems due to wind. Aluminium is stable in air so it may be that this construction method would be satisfactory on a very smooth base.

The most robust of the flexible coverings currently available is butyl rubber. Since the 1950s, artificial rubber sheeting probably has been the most widely used cover. Its advantages are its lower cost (compared to sheet metal or concrete) and the fact that it can be installed over moderately rough surfaces if sharp stones and shrubs are removed. Table 4-5 summarizes the comparative costs of catchment covers.

Over 300 rubber catchments and storage units have been installed on the island state of Hawaii and other Pacific Islands over the past 15 years. The technique involves lining catchments with nylon-reinforced butyl rubber sheeting. This can be done on slopes of up to 40 per cent (Dedrick, 1975).

The capacity of these structures in terms of the volumes of water they can harvest and store ranges from several thousand litres to 5.3 million cubic metres, much of this used for livestock or irrigation. The technique is competitive with other kinds of water provision in both cost and dependability.

Thirty catchments in Hawaii ranging from 1 to 7 hectares in size were

Table 4-5 Estimated catchment and water costs

Material	1,100m² catchment cost $/m²	Probable life Years	Annual cost[a] $/m²	Catchment efficiency (%)	Runoff in 36 cm rainfall zone litres per m²	Water cost $/4000 litres
Butyl, nonreinforced (15 mm)	2.10	10	0.41	95	320	5.15
Butyl, cotton reinforced (20 mm)	2.40	15	0.41	95	320	5.15
Aluminium foil (1 mm)	1.00	10	0.21	80	268	3.15
Polyethylene, black, (1.5 mm)	0.60	3	0.27	90	304	3.55
Polyethylene, black, (6 mm)	0.70	5	0.22	90	304	2.90
Polyethylene, black, (20 mm)	0.90	8	0.21	90	304	2.75
Chlorinated polyethylene, (30 mm)	1.60	5	0.46	100	336	5.50

[a] Includes $0.03 per m² maintenance costs and amortization at 6 per cent interest based on probable life of catchment.

[b] Measured in a 24-cm rainfall zone

Source: Frasier and Myers, 1972

reported in good condition after four years of use. Wind uplift has been minimized by smoothing slopes and weighting the surface with soil-filled butyl bags.

For example, a 1,325 cubic metre livestock reservoir was installed in the area receiving 914 mm of annual rainfall. This structure is capable of harvesting and storing water through a three-month drought.

Problems with rubber catchments are reported to be caused by poor installation, lack of maintenance, poor materials, and damage by animals. Replacement may be required after five to six years. In contrast with these problems are ease of transport and installation; also studies are available on the durability of butyl rubber sheeting (Dedrick, 1973 and 1976). This material will be discussed further in the following chapter dealing with storage.

CHAPTER 5

Storage and Utilization of Rainwater from Ground Catchments

PRIMARY STORAGE

THERE ARE many ways to store water. The storage requirements for rainwater harvesting schemes used to supply water to communities present no new technical problems. The conventional methods of construction, using steel or concrete, can accommodate a great range of storage capacities. Such conventional techniques as metal or concrete tanks tend to be relatively expensive initially, but they greatly reduce seepage and evaporation and have a long life.

If the stored water is primarily for use in agriculture often it is simply stored in earthen dams. Such dams are a ubiquitous feature of the rural landscape. Their capabilities vary enormously depending on environmental and economic circumstances, however. Volumes of 10^6 to 10^7 l are typical.

Depending on the natural topography, earthen dams can be located in existing stream beds and valleys, or they can be formed by extensive excavation. An example of the former, a dam formed in a valley in the Mount Lofty Ranges in South Australia, is shown in figure 5-1. Often such a dam actually enhances the local environment. An example of the latter type where extensive excavation was needed is shown in figure 5-2. This dam is situated on the plain of central New South Wales. It is important in this case for the excavation to be sufficiently deep to keep down evaporation losses relative to storage volume. If this is not done, waste of capital investment can occur, as shown in figure 5-3, including: a windmill, a stone-wall distribution tank and the actual shallow dam itself, which is located on the wide plain between the Mount Lofty Ranges and the Murray River in South Australia.

Care needs to be taken during the construction of retaining walls to ensure that permeable layers of subsoil are removed before building up the structure. Similarly, provision needs to be made to allow for overflow and to ensure that the dam cannot be weakened by erosion, a factor which can lead to catastrophic failure. This is exemplified in figures 5-4 and 5-5. The scenes on the eastern slopes of the Mount Lofty Ranges in South Australia are two aspects of a burst dam formerly holding almost 10^7 l. The retaining walls were weakened by rabbit warrens. Rising water escaped following a thunderstorm,

Figure 5-1 Valley dams in the Mt. Lofty Range in South Australia

Figure 5-2 An extensively excavated pair of dams situated on the central New South Wales plain. The windmill is employed to pump water to a raised supply tank adjacent to nearby farm buildings

Figure 5-3 A too-shallow dam site on the plain between the Mt. Lofty Range and the Murray River in South Australia, where the annual loss by evaporation is approximately five times the mean rainfall. The windmill and stone-walled distribution tank represent a sadly wasted investment

Figure 5-4 A breached earthen dam on the eastern slopes of the Mt. Lofty Range in South Australia. The main wall of this formerly large storage of almost 10^6 l was tunnelled through by rabbits, thereby initiating an erosive flow of water which ultimately led to catastrophic failure

Figure 5-5 A view through the breach of the same dam shown in Figure 5-4, showing the upper settling pond designed to dissipate the energy of water rushing down a concrete channel which was harvested by an extensive ditch contouring the neighbouring slopes. A similar ramped channel, intended to protect the main storage basin, is also revealed

scouring a massively eroded channel which is still a landmark after 35 years (figure 5-6). Earthen dams therefore need regular inspection to ensure structural soundness. Through its failure, the dam shown in figure 5-6 revealed a lack of carefully designed construction. It should have been equipped with overflow regulation in order to dissipate excess water. A simple approach to this problem is shown in figure 5-7, where the rear location of an overflow channel ensures the protection of the built-up sections of the dam.

The water stored in these relatively economical dams comes from local surface runoff, thus water quality cannot be assured. The maintenance of adequate vegetation and ground cover does reduce sediment transport where possible. In high rainfall areas dams form an integral part of pasture management practices in sloping terrain and there it is essential to minimize soil erosion by stormwater. Figure 5-8 shows a typical location of dams beneficial to surrounding pasture maintenance.

Small village tanks, formed by building an embankment or by digging, are used widely in the Middle East, to collect rainwater. The typical storage capacity of such tanks in Iran is 5,000 to 10,000 cubic metres per village. These tanks work in a manner similar to irrigation tanks and face the same problems.

Figure 5-6 The extensively eroded channel which resulted from the scouring flow of escaping water from the failed dam shown previously. This landmark has remained after 35 years, while a modern reinforced concrete tank stands sentinel on the distant hillside

Figure 5-7 A shallow valley dam in southern New South Wales, showing the rear over-flow channel which offers simple protection to the main structure during major rainstorms

Figure 5-8 A pair of dams which simultaneously store water and control erosion of pastures on the Pacific slope of the Great Dividing Range in southern New South Wales

UTILIZATION AND DISTRIBUTION OF WATER FROM EARTHEN DAMS

Water stored in earthen dams is usually regarded as being more suitable for agricultural than domestic purposes. In areas without communal or state-operated irrigation schemes, such dams must be linked to a pumped reticulation system geared to the requirements of the particular crop and terrain in order to be of real use. In some locations inexpensive gravity-fed systems are possible, but not common.

Suspended sediments constitute the chief visible impurity in this water and filtration is rarely contemplated or necessary. However, in order to obviate difficulties caused by an intake of mud into the piping system, most users of earthen dams employ a simple outflow system similar to that shown in figure 5-9. As this diagram indicates, a surface float attached to a flexible outlet tube prevents sedimentary material from being drawn off, at least until the very last reserves are drawn upon.

Where water is precious, as in the occasionally arid environment of South Australia's Eyre Peninsula, farm animals are not given direct access to the storage reservoir, but are only allowed to drink from troughs, the supply being controlled by float-valves (figure 5-10). However, in many other areas of Australia and around the world, sheep and cattle are permitted to drink from storage facilities. The resultant loss of water quality in these areas should be

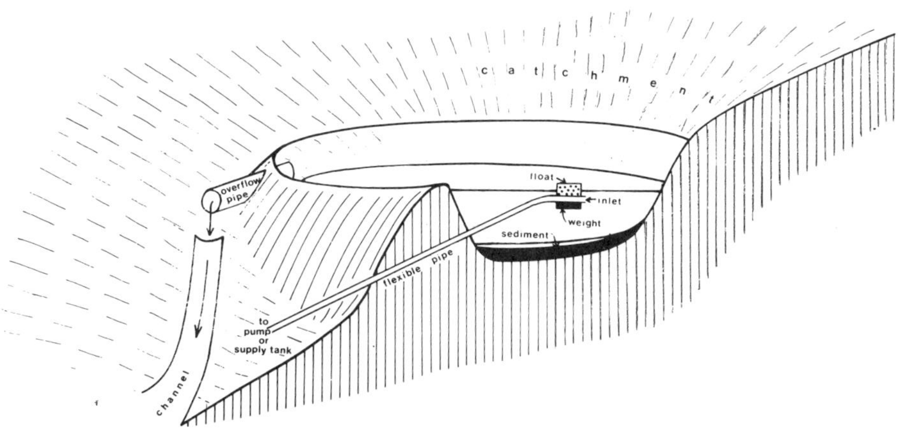

Figure 5-9 Diagram of a typical earthen hillside dam, shown in cross-sectional perspective, revealing a simple, float-controlled outflow system which always selects water from just below the surface, as far as possible from muddy sediments. In some regions, unacceptably high soil porosity may necessitate the installation of an embedded membrane of plastic sheeting. The latter, however, can be damaged both by uncontrolled animals and burrowing freshwater crustaceans

Figure 5-10 A drought-scene on the Eyre Peninsula of South Australia, where sheep search for fodder and are offered water only at certain times (at the trough seen to the left of centre) in order to minimize wastage and loss by evaporation

assessed in the light of savings in costs of pumping and distribution. Unfortunately, the increased opacity of water caused by the additional suspended sediments stirred by trampling animals results in increased absorption of solar radiation in the near-surface levels of the water. This results in an increase in temperature and hence in evaporation rate. Such losses are additional to those caused directly by animals through splashing and spreading of water on surrounding dry surfaces.

SEALING THE BOTTOM OF RAINWATER STORAGE BASINS

Graziers in most central and southern areas of Australia, if they rely on rainfall for water supplies, need to store their requirements for up to 150 days. Since the average grazing animal may be regarded as requiring a volume in the order of 10 l per day, after a conservative allowance is made for losses by seepage and evaporation, an annual storage of 10^6 l is found to suffice for no more than 500 animals! In some regions, where water is reticulated by the State, the cost of such a volume of water would be of the order of $200. The earthen farm dam therefore represents an economically attractive solution which allows costs to be kept down but often suffers from considerable loss of water by seepage and evaporation.

In areas where soils are shallow and where excavation may expose highly permeable subgrade materials, sealing is essential. A wide variety of materials can be employed as sealants. Compacted earth is commonly used, especially in areas where soils have a high clay content. Water and compaction equipment are needed for the construction of compacted layers. These must be at least 20 cm thick.

When added to highly aggregated and porous soils, soluble salts, such as sodium carbonate, sodium chloride, and various sodium phosphates are quite effective in reducing permeability. Salt is broadcast over the soil surface at a rate dependent on the clay content in the soil. Then the salt is disked or harrowed into the soil. Compaction is not necessary, though it does increase effectiveness (Dutt and McCreary, 1975). Although retreatment every two to three years is necessary for continued seepage control, experiments demonstrate that daily seepage loss on salt-treated ponds and tanks can be as low as 2.5 mm or 3.8 mm (the latter from a 125 mm rainfall area).

Other materials such as waxes, asphalt, resin and polymers have been used as soil sealants. Success in maintaining a good seepage with these has been very uneven and thus, at present, these methods are not always well recommended (Dedrick, 1975).

One of the most promising, sodium bentonite, has been found effective for four years. Sodium bentonite reduces seepage in soils containing high percentages of coarse-textured particles. Deterioration of the seal occurs if the stored water is high in exchangeable calcium and magnesium. Powdered bentonite is spread over a dry surface at a minimum rate of 4.9 kg per square metre; then it is disked and compacted. Soils with higher sand content require up to 19.6 kg per square metre (Boyer and Cluff, 1972; Dedrick, 1975).

Membranes and films are used from two broad categories. First are those which retain their attributes when exposed to weathering. Second are those which deteriorate when exposed and therefore must be buried or protected in some fashion. Weather-resistant membranes include asphalt layered with fibreglass or polypropylene and synthetic rubber. Other plastic films, such as polyvinyl chloride, polyethylene, and chlorinated polyethylene, are not weather-resistant and must be buried. Exposed linings are preferable as they permit steeper embankments. In most cases, the excavated area must be cleared of sharp objects, and in some cases cushioned, in order to accommodate puncturable lining materials. Soil sterilants are recommended to prevent puncture by plant growth.

Asphalt materials include: catalytically-blown asphalt, asphalt-saturated felt, hot-sprayed asphalt, and asphalt cement with crumb rubber. Catalytically-blown asphalt reinforced with asbestos fibre is most effective when buried.

Asphalt-saturated felt, a prefabricated liner, has the advantage of easy instalment and maintenance; but it has problems of leakage through the joints. Hot asphalt can be sprayed on the felt as a sealant, eliminating the joints, but this requires special spraying equipment (Dedrick, 1975).

Recent experiments combining asphalt cement with crumb rubber obtained from discarded tyres, show that when this material is sprayed as a liner, it is less costly and is still as effective as catalytically-blown asphalt. However, the asphalt/crumb rubber material must be covered with chips or soil (Froebel *et al.,* 1978).

Asphalt can also be used as a waterproofer over a substrate matting of fibreglass or polypropylene, with new seal coats added over time. No measurable seepage has been found in ponds using this sealant. Moreover, this technique is good on rugged terrain, since it requires no heavy equipment for installation, and the membrane can be made to conform to surface irregularities, including partially exposed boulders (Fink and Cooley, 1973).

Buried plastic films (between 0.02 and 0.03 mm thick) of polyvinyl chloride, polyethylene and clorinated polyethylene are effective seepage barriers, but are prone to problems such as burrowing animals or vandalism. Recent research in Arizona tested the performance of several plastic-lined tanks. A 455,000 litre tank for storing harvested water developed leaks due to burrowing animals and had to be covered with a soil-blanketed layer of used tyres. Another smaller plastic-lined tank which developed leaks was covered with a 13 mm layer of wire-reinforced concrete mortar, a trouble-free seepage control material. Several polyethylene-lined tanks were lined with a soil-covered layer of used tyres which acted as a cushion, and then filled with rocks. This is a deterrent to vandalism which moreover reduces losses to evaporation by up to 90 per cent. In this instance, storage capacities are greatly reduced. The resulting tanks are virtually indestructible; they may, however, be fairly costly (Dedrick, 1975; Cluff *et al.,* 1972).

Investigations and prototype construction of low-cost rainwater reservoirs in Sudan have produced several novel possibilities (Anon., 1967). The emphasis was on "intermediate technology"—or cheap and easy construction—using large amounts of relatively cheap labour and minimum amounts of imported and expensive materials. Excavated tanks were supplied with rainwater in an area having about 400 mm of annual precipitation. One model reservoir had a lining consisting of four layers of polythene sheeting, with layers of mud between the sheets, so that any hole in the sandwich liner would tend to plug itself with mud carried down by escaping water.

Newer storage techniques involve the use of flexible surface coverings, as discussed in chapter 4 dealing with catchments. Polyvinyl chloride or polyethylene sheets are less vulnerable to wind or sunlight when used as reservoir liners but they are still not as satisfactory as butyl rubber. Synthetic rubber membranes such as butyl rubber and ethylene propylene diene monomer, can be used as linings since they are resistant to weathering. However, they must be protected from puncturing and damage by animals.

In Hawaii liners consisting of top and bottom layers of butyl, laminated 0.8 mm nylon, have recently been used to line stock reservoirs.

COVERING THE SURFACE OF RAINWATER STORAGE BASINS

A basic question associated with the design of a reservoir for rainwater is whether or not it should be covered. One possible justification for covering a reservoir is to prevent the loss of water through evaporation. In order to justify the covering of a reservoir on these grounds there should be an estimate of both the potential evaporation loss and the cost of compensating for such loss by building a slightly larger catchment area and storage volume.

The covering of a reservoir containing water for domestic use is always desirable on grounds other than the reduction of losses. Control of the quality of the water is probably the principal concern. For example, algae are normally encountered in surface water supplies exposed to sunlight. Their accumulation in a reservoir can cause taste and odour nuisances as well as being aesthetically objectionable. Algae require nutrients and sunlight to survive. The open reservoir of a rainwater harvesting scheme provides the sunlight requirement. However, the quantity of nutrients in harvested rainwater would generally be much less than in a conventional scheme (depending on the catchment area). Pure rainwater would wash only a negligible amount of organic matter into a reservoir from an impervious and protected catchment designed specifically for a drinking water supply scheme. The question of possible algae problems and the justification for eliminating them warrants consideration in each particular situation.

Pollution of the water in an open reservoir could result from wind-blown particles or from bird life. Near oceans seabirds often seek quiet inland waters in times of storms. If scavenger birds are visitors to the reservoir the danger of contamination of the harvested rainwater exists. To justify covering a reservoir may be difficult in quantifiable terms but the prevention of pollution will always provide a strong reason for such construction.

In many areas (for example, Singapore) local ordinances prohibit any open water surface because they provide breeding places for vectors, such as mosquitoes, which can be carriers of malaria and yellow fever. Where such situations exist the covering of the reservoir is not a subject for discussion: it must be covered. Where such public health considerations do not apply it may still be advisable to cover the reservoir simply to eliminate a breeding area for nuisance insects.

Covering of reservoirs by traditional methods of rigid construction is generally quite expensive, particularly for large reservoirs. Recent developments, however, allow consideration of more economic means especially for evaporation control.

From 1950 to 1970, most evaporation control research focused on monomolecular layers, films, and long-chain alkanols. These have proved less effective than anticipated because they do not reduce incoming solar energy, and moreover, are vulnerable to wind. Long-term field studies demonstrated that these covers suceeded in reducing evaporation by only 20 per cent.

A more recent approach has concentrated on reducing the energy available for evaporation, either by reducing the amount of solar energy entering the stored water or by reducing the transport of water vapour above the water surface. The most effective of these methods, summarized below, is initially more expensive than monomolecular layers, but in the long run is durable and efficient. Table 5-1 summarizes the levels of evaporation reduction achieved by various energy-reducing methods.

Attempts to reduce evaporation by dyeing water a lighter colour have not been particularly successful to date. Wind barriers have not been researched in detail, but one study indicates that wind baffles do not reduce evaporation significantly. Shading the water surface with plastic sheeting has been a more successful evaporation retardant, but there are cost problems with the construction of large-scale support structures, and problems with strain and wind damage to the supported shade-material.

Floating water covers, the most widely researched evaporation control method to date, exhibit effective results, ease of use, and low maintenance requirements. These covers act both as reflectors and as vapour barriers. The covers range from small individual particles, such as perlite ore, polystyrene beads, and wax blocks, to larger pieces, such as polystyrene sheets, rafts and butyl sheets, and complete one-piece covers, such as continuous wax covers. Table 5-1 compares results obtained with these methods, but it must be remembered that research was conducted under greatly varying conditions.

Of these methods it was found that continuous wax, polystyrene rafts, and butyl rubber are the most readily available and the least difficult to instal. For the continuous wax method paraffin wax is used, like that used for canning, which melts at 53° to 54° C and forms a continuous cover during summer months. This wax can either be placed on the surface as blocks which will later be melted by the sun to form a wax layer (about 3 mm thick) or melted with a heater and sprayed or poured on the water.

Polystyrene rafts are constructed of 1.2 × 1.2 m sheets of expanded polystyrene, 25 mm thick, coated with emulsified asphalt and covered with a layer of chips. They are then coupled together using a clamp made of PVC pipe. An outer frame of 32 mm diameter PVC pipe is used as a bumper for the rafts. Continuous covers of low-density, closed-cell synthetic rubber sheeting, available as 1.2 m wide roll stock, have been fabricated for use on water storage tanks. Covers have been fabricated from material 5 and 6 mm thick.

All three covers — continuous paraffin wax, polystyrene rafts, and foamed rubber—reduce evaporation by 85 to 95 per cent. The cost of water saved in high evaporation areas compares favourably with alternate water sources (Cooley *et al.,* 1975).

A comparison of the savings using these methods based on transporting water is provided in table 5-2. Wax cover would result in a saving of $55 per 50 head of cattle per year.

Another technique for reducing evaporation involves minimizing the surface-area-to-volume ratio by utilizing a compartmented reservoir with a pump to keep the water concentrated, minizing its exposure to the atmosphere.

Table 5-1 Evaporation reduction achieved by various energy-reducing methods

Method	Area of water surface covered %	Evaporation reduction %
1) Changing the water colour:		
Dye in water	100	6-9
Shallow, coloured pans	100	35-50[a]
2) Using wind barriers:		
Baffles	-	11
3) Shading the water surface:		
Plastic mesh	47	44
Blue polylaminated plastic sheeting	100	90
4) Floating reflective covers:		
Perlite ore	78	19
Polystyrene beads	78	39
Wax blocks	78	64
White spheres	78	78
White butyl sheets	86	77
Polystyrene sheets	80	79
Polystyrene rafts	100	95
Continuous wax	100	87
Foamed butyl rubber	95	90

[a] Evaporation from white pan compared with that from black pan.

Source: Cooley, 1975.

Table 5-2 Cost of transporting water compared to evaporation reduction. Example for central Arizona (USA) for a 3-month period

WATER REQUIRED:			
(a) Grazing capacity for 50 cattle			7 ha / head / mo
(b) Water needed 45 l / head / day			207,000 l
EVAPORATION LOSS:			
(a) May, June and July			0.88 cm / day
(b) Stock tank (46m² area)			570 l / day
(c) Total Evaporation (97 days)			55,300 l
COST OF TRANSPORT:			
(a) Round trip			12.8 km
(b) Tank truck capacity (assume one round trip per hour)			2,300 l
(c) Round trip cost: Petrol	= $ 0.50		
Maintenance	= 1.50		
Driver and water	= 3.00		
			Total: $ 5.00
IF EVAPORATION REDUCTION COVER:			
(a) Water required: Cattle	207,000		
Evaporation loss	5,300	(save 90% of 55,300)	
Total	212,300		
(b) Cost if transported: $\frac{212{,}300 \times \$5}{2{,}300}$		$ 465	
COST OF COVERS (46 m²):			
Wax	$ 50		
Gravel-covered polyethylene rafts	80		
Foamed butyl rubber	125		
SAVINGS:			
Difference in transport cost			$570 - 465 = $105
Savings using wax cover			$105 - 50 = $ 55
Savings using gravel-covered polyethylene rafts			$105 - 80 = $ 25
Foamed butyl rubber—no initial savings			$105 - 125 = -$20

Sources: Cooley, 1975; Cluff, 1978; Evans *et al.,* 1975; Fairbourn, 1975.

Finally, evaporation can be reduced by filling reservoirs with sand or rock. This actually is not a novel technique at all; it has been used for centuries in various parts of the world. This technique reduces evaporation at the cost of reducing storage capacity.

Most sand-filled reservoirs are built in stages, each stage filling naturally with stream-carried sand. Additionally, these reservoirs are caused when dams are constructed across intermittent, sand-carrying streams. Weed growth must be controlled, and a moderately steep gradient with a large supply of sand is required for success with this method.

The versatility of butyl rubber permits novel types of rainwater reservoirs to be considered. The simplest type is a "rain trap" installation for supplying water to livestock. This consists of a sheet of butyl spread on the ground as a catchment. It supplies rainwater to the butyl storage bag lying downhill. Water for livestock is supplied from the storage bag to a trough through an automatic float valve. Prefabricated bags for this application are available, made of nylon-reinforced butyl sheeting. The one-piece bags are prefabricated with built-in inlet, outlet and overflow pipes. They are more costly and heavier than two-piece bags which can be constructed simply in the field. The two-piece bags are assembled by cementing a pit liner sheet and a cover sheet together with pipes where convenient (Cooley, 1975; Dedrick, 1975). The one-piece bags are available commercially in sizes up to 228,000 litres. Two-piece bag sheets can be made to any site specifications, with on-site splicing a further possibility.

The butyl bag may be used inside a concrete basin and any impermeable catchment surface.

In Kenya the supplier of butyl has designed a prototype where a cylindrical cage made of galvanized steel wire is used, such as that used for fencing or for reinforcing steel for concrete. The light steel frame provides the structural support while the butyl bag provides the watertight storage.

A comparative study of several water storage systems using closed synthetic rubber bags and plastic or rubber-lined pits with an average capacity of 114,000 litres was carried out in Utah over an eleven-year period. The systems generally worked well over the first seven to eight years, and after that exhibited failures owing to a variety of causes. The most common problems included: damage by livestock and other animals; snow accumulation on the top of the storage bags; lack of maintenance, design mistakes including underestimation of water requirements, poor site selection, lack of reliable precipitation, and high rates of evaporation (Dedrick 1976; Dedrick and Williamson, 1973).

In southeastern Arizona several types of experimental catchment and butyl bag storage facilities were installed in an area receiving annual rainfall of 150 to 400 mm and experiencing summer temperatures up to 43° C and winter lows of −4° C (McBride and Shiflet, 1975).

The researchers discovered that problems arise when the varying water level in the bag warps the rubber inlet and overflow pipes. This problem can be corrected by relocating the pipes in the lower half of the bag. Another problem arises when filled bags tear out their supports and roll towards the downslope side of their container pits, thus breaking their plastic outlet pipes.

The study concluded that rock slopes, slick rock areas, and ledges may be superior to rubber, steel, and fibreglass both in terms of water collection and costs.

LOCATION OF WATER SUPPLY FACILITIES

For rainwater harvesting schemes based on catchments which depend on rocks or other topographic features there is little choice concerning the location of the facilities. The catchment area is defined independently from the requirements for water supply. The design problem is to determine the dependable capacity of a supply based on the available catchment and to select the reservoir size needed to develop as much of the capacity as is required. The reservoir location, between the outlet of the catchment area and the demand centre, must be selected to minimize pumping and transmission costs. Under the most favourable circumstances water would flow to all points of use entirely by gravity.

Where a natural catchment such as a roof or rock surface does not exist, the site of all components of the precipitation harvesting scheme can be rationally selected principally on the basis of the distribution of the demand for water. The cost of land is the other major criterion in site selection. The opportunity to locate all components of such water supplies close to their point of use distinguishes them from conventional schemes, whose location is generally controlled by topography (surface water schemes) or geology (ground water schemes).

In unsophisticated communities, where water is carried from a central supply to the point of use by the consumers, the matter of site determination becomes a simple transportation problem. There will be an economic limit to the distance between the location of the water storage and the point of use. As water weighs one kg per litre and as people carrying water cannot walk more quickly than 3 to 5 km per hour, the effective radius of the area which such a water supply can serve is quite small. Thus, before the location of the storage facility is finalized many factors should be taken into consideration (Deshmukh, 1979). Storage tanks or bunds on flat ground can be located near the water source and water can be pumped from the tank with low-lift pumps (Geddes, 1968). Storage tanks or bunds can also be located near any of the waterways, or channels which are common dispersal systems (Jaikumar, 1974). Wright (1956) felt that the ideal location for storage ponds is: (a) a natural saucer-shaped area; or (b) a shallow ravine with a small flowing stream or a fairly level area near a large stream where water can be diverted to the pond to augment the seasonal rain harvest. He further emphasized that the storage facility must be located preferably at a site having a high water retention capacity. It has also been reported in project planning for dry land that the location of any storage facility necessarily should be at the lowest point of the watershed.

Von Oppen (1974) felt that the economic aspects of application of stored water should be taken into consideration for fixing the location of the facility. He felt that the lifting and conveyance of collected water to donor areas of the watershed is often a limiting factor due to the prohibitive cost. This is because the high cost of lifting and piping of harvested water increases the cost of water and would negate the value of water harvesting. Jaikumar (1975) suggested that the ratio of an area irrigated to the area donating water should be taken into consideration as another approach for fixing the location of storage facilities. Further, he emphasized that the location of the tanks should be such as to avoid high pumping and conveyance charges. Scientists in India have developed a mathematical model for determining the optimal size and location for a dug tank connected to a gravity irrigation scheme. Saxena (1976) suggested that a number of bunds of suitable capacity be located across a slope suitable for gravity irrigation. This view is in agreement with the system developed at Hyderabad, India, by Jaikumar (1975) where the tank location is such that the same terrace or the next lower terrace can be irrigated with a low lift of about 2 metres.

Geddes (1963) pointed out that the measure of economy is the storage excavation ratio. Any excavated tank or bund on flat ground with an embankment on three sides has the most uneconomical storage because of a 1 : 1 ratio of storage and excavation. The ideal ratio is 6 : 1 and can be obtained in the case of nest type ponds. There are many practical difficulties in accomplishing this due to the inlet arrangement. Hence, the site of the bund has to be related to storage. Much thought has been given to this aspect in India and a mathematical model has been developed for determining the optimal size and location. More storage can be obtained by:

(a) locating a bund in waterways having more slope;
(b) constructing it across the lower side of the catchment; and
(c) making a dam across a gully.

In the case of trapezoidal and cup-shaped bunds the increase in volume is directly proportional to the C/S area.

Proper planning of rainwater harvesting schemes should use an integrated approach in which collection, storage, treatment and use are integrated to achieve maximum efficiency.

WATER QUALITY AND TREATMENT OF WATER FROM GROUND CATCHMENTS

Rainwater harvesting schemes should produce water of higher quality than any other form of surface water supply because the catchment is deliberately constructed, or at least ulitized, to provide water for domestic use. Because most catchment areas are relatively small, possible pollution can be reduced by controlling or eliminating the use of the area for any other purpose but water supply. Fencing would be an essential component of any well-designed rainwater harvesting project. Wind-blown particles will be deposited on the catchment, however, and birds and other creatures may have difficulty in reading man-made warnings of 'keep out'! Complete elimination of pollution over the catchment area is impracticable.

For most rainwater harvesting schemes the control over the entry of toxic substances into the water should not be difficult. Chances of contaminating a well-operated scheme with pathogenic organisms should also be remote, because water, to act as a vehicle for the spread of specific disease, must be contaminated with the disease organism from infected persons (Calvato, 1958), and a fenced catchment (or elevated roof) is unlikely to become contaminated. The storage reservoir should be above ground or, if in an excavation, made watertight to eliminate the possible entry of polluted ground-water. If the reservoir is covered, the possibility of any transmission of disease organisms by insects or birds would also be eliminated while at the same time the growth of algae would be prevented.

Water quality can be positively influenced by the design of a sensible rainwater harvesting scheme. The surface of the catchment determines whether or not the runoff will contain suspended or dissolved solids. Since some debris is liable to accumulate on the catchment or in conduits leading to the reservoir some elementary precautions can reduce pollution of the water in storage. Fine mesh screens on flap valves at all inlets, ducts and spillways will prevent the entry into the reservoirs of animals, birds and large insects as well as debris. Simple maintenance, including the removal of such debris when it accumulates, would prevent blockages in water conduits and the possible waste of water which could result from overflows.

Silt and sediment from most ground catchments would be deposited in the silt trap which should be part of the design. If the catchment area produces runoff containing very fine suspended matter, and if the quiessence of the silt trap and reservoir does not clarify the water sufficiently through sedimentation, some sort of filtration might be advisable. Cloudy water is not particularly attractive, and water which is hygienically safe may nevertheless be unpalatable.

Runoff from an elevated catchment can be easily fed through simple sand or charcoal filters. But if the water is collected at ground level and is stored in an excavated reservoir, it can also be filtered before it is withdrawn. In Sudan, for example, water is removed from "haffirs" or excavated tanks by pumping it from a well shaft in the corner of the tank. This shaft is surrounded by sand through which the water must flow before it is pumped out (Anon, 1967). Such sand filters can be coupled with plastic liners to prevent seepage and the result is a substantial storage facility (figure 5-11).

If precautionary disinfection of the community water supply is required the preferable system for small rainwater harvesting schemes would probably be an unsophisticated type of chlorination. Good discussions on appropriate techniques are available in manuals by Wagner and Lanoix (1959), W.H.O. and the US Public Health Service (USDHEW) (1965).

Treatment of water by individual consumers is also possible. Suspended matter can be removed in diatomaceous earth filters, which are available for home use in the shape of cylindrical jars. Simple charcoal filters can also be devised if the taste of the water needs to be improved. Home disinfection can be accomplished by boiling the water. Chemical disinfection of small quantities of water is also possible, but any decision concerning treatment of water supply of this type should be based on both the water quality expected from the scheme and the consequences of not providing treatment. A designer must also be aware of the education of the consumers. Not all users of water may be relied upon, for example, to utilize disinfectant tables even if they were freely available.

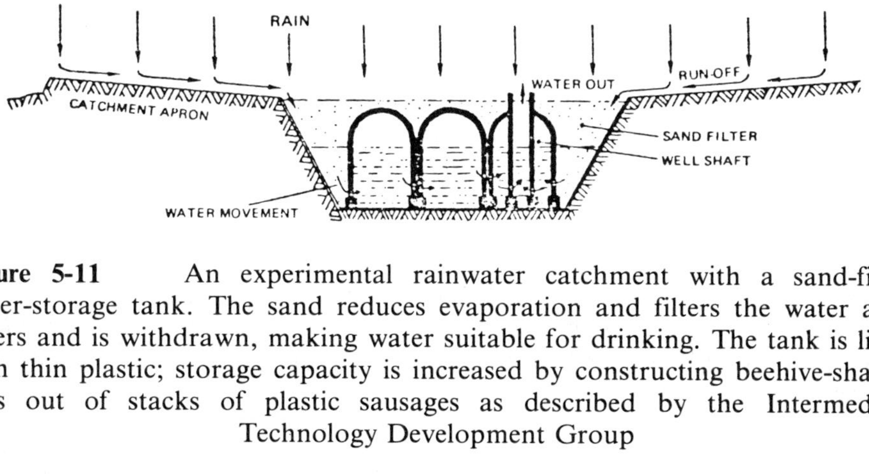

Figure 5-11 An experimental rainwater catchment with a sand-filled water-storage tank. The sand reduces evaporation and filters the water as it enters and is withdrawn, making water suitable for drinking. The tank is lined with thin plastic; storage capacity is increased by constructing beehive-shaped cells out of stacks of plastic sausages as described by the Intermediate Technology Development Group

RAINWATER STORAGE BY UNDERGROUND RECHARGE

This type of water storage can be regarded as a special application of rain harvesting. The controlled yield of rain-fed springs developed in Europe and the Middle East is a very simple and inexpensive method for supplying safe and healthy water to small, rural communities.

The general concept here is simple; for example, water tables slope generally from higher terrain down in the direction of plains. A continuous horizontal groundwater flow develops along the slope, which transports the water infiltrated within the intake area to the lower-lying plains. In the plains the system is either drained by evaporation or the water reaches the surface in small springs which spread out at the edges of the plains.

Horizontal galleries were constructed thousands of years ago to collect this excess infiltration water within the intake area and to bring it to the surface in concentrated form suitable for utilization, instead of allowing it to become groundwater flow. The ancient structures were usually composed of vertical dug wells which collected the water and gently sloping tunnels which channel the water from the head wells to lower points of the terrain. Sometimes the tunnels were supplemented with vertical shafts, which facilitated the removal of the earth during the excavation and ensured the ventilation of the tunnels. Often the head wells were supplemented by laying stoneware drains to increase the water collecting efficiency of the systems. These complex structures are called *qanats* in Iran; *karezes* in Afghanistan and Pakistan; *foggaras* in North Africa; and *falaj* in the Arabian Peninsula. In these areas the ancient construction methods are still used. In practice, a qanat consists of a series of vertical shafts in sloping ground, inter-connected at the bottom by a tunnel whose gradient is flatter than that of the ground (figure 5-12A). The first shaft is usually sunk in a hill slope to a level below the water table, and is called the mother well. The direction of the groundwater flow is determined by evidence provided by the presence of springs, vegetation and swamps in low ground. Shafts are sunk at intervals of 20 to 30 m in a line along the direction so determined. The shafts are rectangular in plan, the typical size being 0.7 × 1 m. Figure 5-13 shows a photograph of a shaft of the Chintala Karez near Kabul. The shaft is lined with stone for a depth of about 1 m from ground level to prevent the edges falling in.

Figure 5-12B shows the actual technique of construction of qanats. A man excavates the shaft, and another man lifts the excavated soil to the ground with the help of a windlass. The soil is dumped around the opening of the shaft to form a small mound, so that runoff from the surrounding land does not enter the shaft bringing silt with it. After the shafts are sunk, they are interconnected by a tunnel whose height is of the order of 1 to 1.5 m and width 0.4 to 0.8 m, just enough for a man to work in. The excavated soil is again transported through the nearest shaft. The man working the tunnel carries a lamp with him. If the soil is firm, no lining is provided for the tunnel. In loose

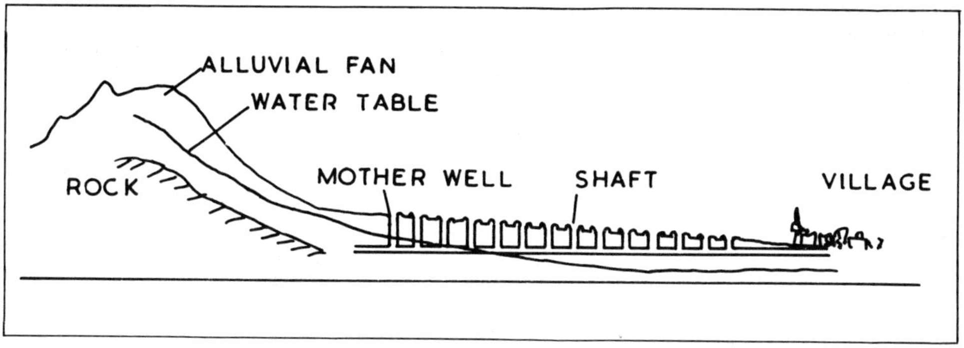

Figure 5-12 A qanat **(above)** and its construction **(below)** (Krosigk, 1971)

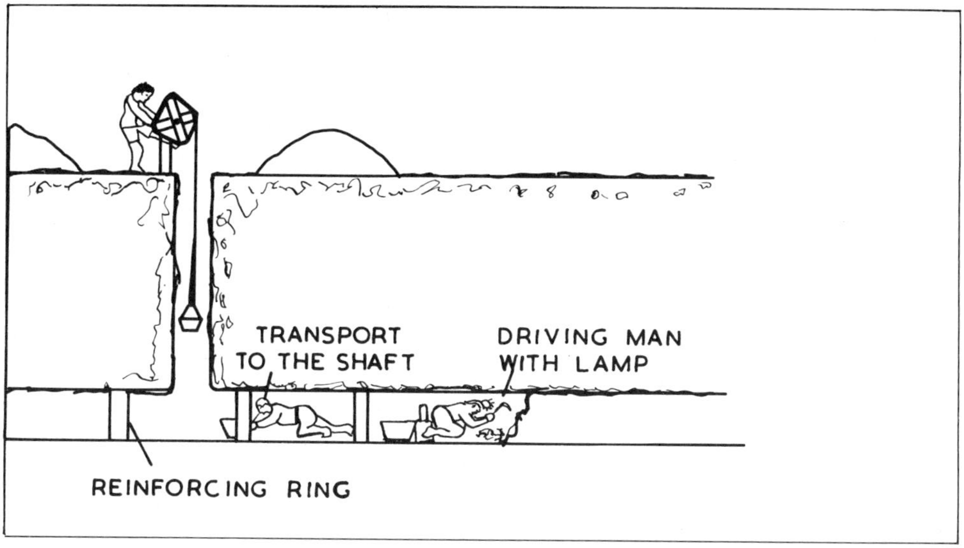

soil, reinforcing rings must be installed at intervals in the tunnel to prevent its collapse. These rings are usually made of burnt clay. The slope of the tunnel is adjusted to keep the depth of flowing water constant (Krosigk, 1971). When the qanat is in operation, the shafts serve to provide aeration to the flowing water as well as entry for maintenance work.

Qanats are found not only in Iran and Afghanistan, but in China, Pakistan, the Syrian desert, the Arab Peninsula, Palestine, North Africa (Sahara), Spain, Chile, Peru and Mexico (Evenari *et al.,* 1971). Megasthenes, who was travelling in India in the year 300 BC records the existence of qanats in Baluchistan, now a province of Pakistan. In the Middle East, qanat building is a skill in which some families specialize and which provides their livelihood.

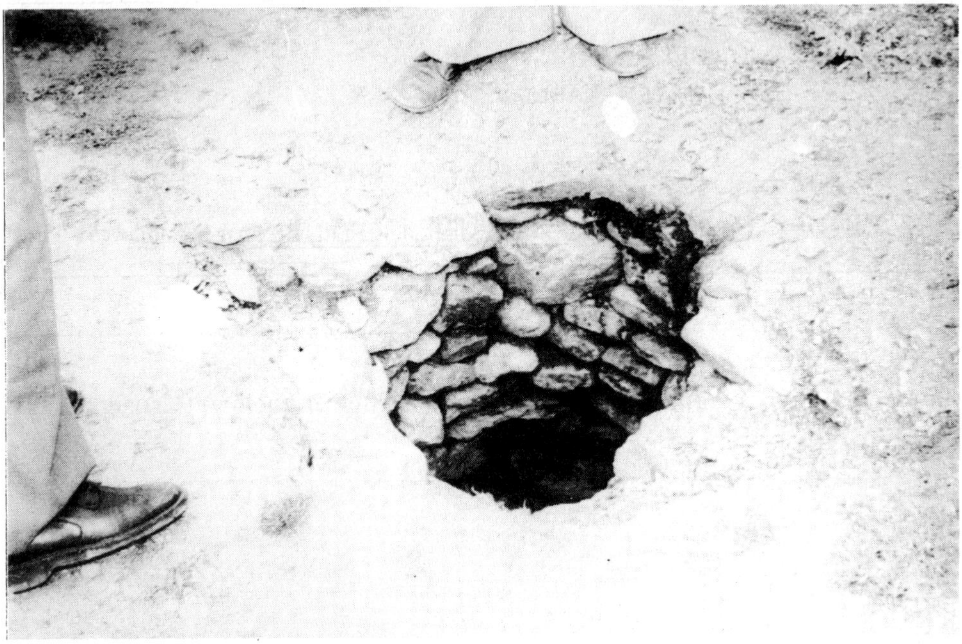

Figure 5-13 Shaft of qanat near Kabul

Qanats are built as a rule in winter when the demand for oxygen by the workers in the tunnel is smaller than in summer (the oil lamp also needs oxygen, though at the present time battery-operated lamps are used in many places). Building a qanat is also somewhat dangerous, since there is always the risk of a cave-in, and asphyxiation can result in the narrow tunnel. In the light of these considerations, most qanats are built by younger people (Gruner, 1944).

Figure 5-14 shows the outlet of a qanat near Kabul. The water flows in a channel to irrigate lower land. When the water is intended for consumption in a village, an enclosure is erected to cover a ground level cistern into which the qanat discharges, so that pollution is avoided. Water is then taken in pots from the cistern. It is said that the city of Tehran got its water till 1933 from qanats excavated from the Elburz mountains. A qanat may take two to three years to build, and costs about US$2,250 per km length. The maximum discharge from a qanat is of the order of 30 litres/second.

A modern form of these structures was developed in Hungary by combining horizontal water-collecting galleries with groundwater storage and utilizing rain harvesting in this way. The most characteristic examples can be found along the northern shore of Lake Balaton. Here springs rising in the valleys had provided the summer resorts located nearby with healthy drinking water until the demand exceeded the amount available. An increase in the yield of the springs was possible by constructing horizontal galleries but this was not

Figure 5-14 Outlet of qanat near Kabul

sufficient, since the seasonal fluctuation in demand is doubled from June to August. Natural discharge is, of course, highest in March and lowers gradually till September. Thus demand cannot be covered from natural sources in the summer months even if the annual sums of consumption and discharge respectively are equal.

A solution was found by regulating the discharge at the outflow of the gallery. The storage capacity of the aquifer is depleted by lowering the discharge level, and thus the water table, during the period of high consumption. Thus a greater amount can be provided for utilization than the natural yield. When the season is over and the demand decreases, the level at the outflow is raised gradually to its original elevation by closing the valves regulating the outflow and the subsurface reservoir is refilled in this way. The low-water table at the beginning of the accumulation period strengthens the infiltration, which retains a part of natural runoff within the area. This method is thus a special type of rain harvesting.

The actual example chosen here for illustration is the structure now in use in the Nosztori Valley in Csopak, Hungary. In general, the valley is underlain by a geological structure which provides the basis for local, rainfed springs. The water-collecting gallery was controlled by means of a siphon which allows discharge into a regulating main. Figure 5-15 shows the construction of the siphon.

In India similar principles lie behind the use of percolation tanks and percolation canals (*rapats*). The general idea is to allow rainwater to percolate

Figure 5-15 The construction of the siphon

into a ground area where some regulation is achieved by natural underground impervious layers or concrete (or stone) retaining walls or casing. Figure 5-16 shows a cross-section of such an impervious layer. The details of the inner wall can be seen in figure 5-17. It is interesting to note that the technique of using percolation tanks and rapats was accidentally rediscovered in Israel when an attempt was made in 1958 to create a storage reservoir by building a dam across the Shiqma Wadi near the Gaza strip border in the southern coastal plain at Karmiya (Orni and Efrat, 1964). The area is characterized by coarse, pervious sand dunes. The reservoir failed to function satisfactorily as a storage facility because the water percolated rapidly into the ground. It was, however, noticed that wells downstream of the dam showed a rise in the water level. It is estimated that this dam replenishes the groundwater to the extent of 3.5 million m^3 per year. Two more experimental percolation reservoirs at En Kerem and Bet Netofa have been built subsequently.

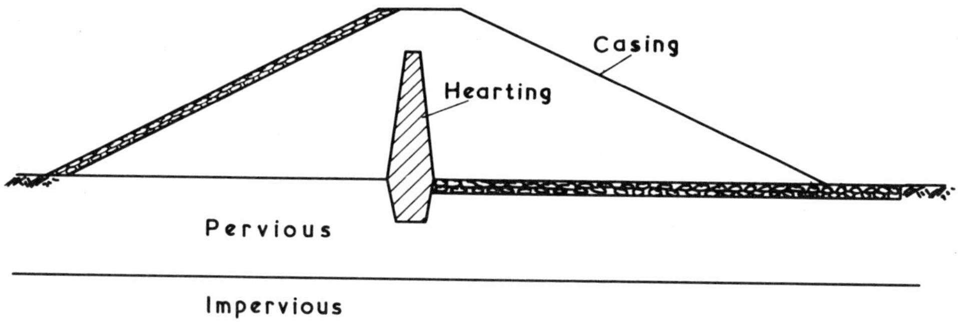

Figure 5-16 Cross section of a percolation tank bund in India

Figure 5-17 Percolation tank near Pune in India

CHAPTER 6

The Ground Catchment on Manda Island

INTRODUCTION

IN THE previous chapters, we have discussed the range of materials and techniques available for harvesting and storing rain and stormwater. In this chapter we would like to show an actual case where people in a rural area had to be provided with a water harvesting scheme. As a practical exercise it shows what problems will arise and how they can be resolved in similar schemes.

Manda Island, the location of the scheme, is a small coral island, located off the coast of Kenya not far from a similar island, Lamu, which is a tourist resort.

No surface or groundwater sources exist on Manda Island. The island is generally low-lying, with most of its surface (approx. 16 km) not more than 17 metres above sea-level. Due to the topography and geology of the island, rainwater does not collect in any developed drainage pattern: some of it sinks into the coral formation as soon as it falls and what is left tends to collect in small pools and either evaporates or slowly infiltrates the coral.

The adjacent Lamu Island has no surface water either but does have a good source of groundwater. Much of Lamu Island is sandy, the large sand dunes on the seaward side of the island being the location of the 20 wells from which water is pumped to provide the public water supply to the people in the town of Lamu. The capacity of this aquifer was unknown but was not believed to be much greater than the withdrawal rate in 1969, approximately 120,000 litres per day (lpd).

The water supply for Manda Island could not be provided by water from Lamu Island. No other sources of surface water or groundwater on the mainland or nearby islands are assumed to be available to supply Manda. As desalination of seawater on such a small scale in such a place would be uneconomical, there are only two possible means of providing the required water supply:

(a) importation of fresh water by ship; or
(b) rainwater harvesting scheme built on Manda Island to supply local requirements.

Costs for the first alternative can be estimated. The nearest place to Manda Island where a ship could obtain supplies of fresh water is Malindi, 128 km to the south. The water would have to be purchased from the government-operated Malindi water supply. The source of Malindi's water is the Sabaki River. Water pumped from this river must be treated by coagulation, sedimentation, rapid sand filtration and chlorination before being pumped to the town of Malindi. The retail price of this water to consumers is $0.90 per 4,000 litres.

To provide 30,000 litres of water daily on a 256 km round trip would necessitate leasing one small ship or barge continuously. The transportation costs would be the largest element in the total cost but initial purchase of the water and storage and distribution of the water at Manda Island would contribute to the costs. A preliminary estimate of the costs involved in this alternative is about $200 per day ($73,000 per year).

The water supply for Manda Island could be provided more economically by a rainwater harvesting scheme than by any other means.

The per capita consumption of 23 lpd used as the basis for design of the water system is quite low. Habbil and Doland (1975) have suggested that the minimum human requirement for biological purposes has been estimated at one lpd but the normal estimates of minimum water requirements in developed countries are usually higher.

On Manda Island no water would be used for waterborne sanitation. Most of the water requirements would be for washing, cooking and drinking. To determine the level of water demand in these circumstances a survey was carried out on Manda Island over a six-day period in August 1968. At that time, many of the 114 families on the island obtained water from one small rainwater harvesting scheme there. This water was removed from the 80,000 litre reservoir by a hand pump and carried in pails to the farmers' buildings. Most people carried the pails on their heads (16 litres or 18 kg) but several used a donkey for the task. The per capita consumption measured during the survey period was 14 lpd. This figure is probably less than the actual average consumption since auxiliary supplies of water were believed to have been imported during the same period from Lamu Island.

Farms were laid out on a grid system, with each farm facing on to a 30 m road. Farmers live on their plots, presumably near the road. The community centre was located at the crossroads in the middle of the farming area.

One water source (supply No. 1) was located in the community centre to serve the daily demand there of 2,000 lpd. At least two more sources (supply Nos. 2 and 3) were required in the area.

The existence of three water source sites would minimize the total distance between all consumers and their nearest water source, with no plot on the settlement more than 1.6 km from water. Water supplies No. 1 and No. 3 would each serve 80 farms, so each of these schemes would make a demand of 11,300 lpd. Water supply No. 2, serving 40 farms and the community centre would have a demand of 7,700 lpd.

The rest of this chapter concentrates on the design of a rainwater harvesting scheme for an average demand of 10,800 lpd. A different scheme for the smaller demand of water supply No. 2 was designed using the principles developed, but three similar schemes were assumed to be needed, having a total capacity of 32,400 lpd. This was to give the settlement an excess supply capacity of 3,200 lpd or just about 10 per cent of the total estimated demand of 30,300 lpd (table 6-1).

Table 6-1 Summary of water needs on Manda Island

	litres per day
Demand:	
Per capita consumption (observed)	14
Per capita consumption (design figure)	23
Families = 114	
Population used in design = 1,300	
Total demand expected	30,000
Supply Needed:	
Supply No. 1	11,300
Supply No. 2	7,700
Supply No. 3	11,300
Total Supply	30,300
Design:	
Scheme No. 1	10,800
2 Schemes later on	21,600
Total design	32,400

LOCAL CLIMATE

Characteristics of the local climate are indicated in table 6-2. The coefficient of variation for the precipitation data (calculated according to Chow, 1964) is 0.3. Arid regions have coefficients as high as 0.5: for well-watered regions the value is as low as 0.1. Using this criterion, Manda Island is semi-arid. This concurs with Meigh's classification (1968) for the region, which is based on water requirements for plants and the deficit of precipitation in relation to the potential evapotranspiration of the area.

Table 6-2 The local climate on Manda Island

Minimum monthly mean air temperature :	25.1°C
Maximum monthly mean air temperature :	28.1°C
Average total sunshine hours per year :	3,242
Average relative humidity at 3 p.m. :	75%
Mean annual precipitation :	905 mm
Average number of days precipitation per year :	85 mm
Mean annual potential evaporation	2,327 mm

There is a distinctly seasonal pattern to local precipitation as indicated in table 6.3.

Table 6-3 Monthly mean rainfall

	(cm)
January	0.5
February	0.3
March	2.0
April	12.5
May	31.8
June	15.5
July	7.5
August	4.0
September	4.1
October	4.1
November	4.0
December	2.8
Total	89.1

Nil precipitation was reported in almost 20 per cent of the months but never in the period from April to July.

To test whether or not the precipitation data are statistically reliable is desirable but difficult for records from a single station. Fortunately, rain gauges tend to under-register the amount of precipitation due to surface wetting of the instrument, leakage, and evaporation, thus the design of the rainwater harvesting scheme proceeded on the assumption that the available records provided an accurate indication of local conditions.

CONSTRUCTION METHODS

For a rain harvesting scheme the basic design problem was to find the minimum cost for producing a sufficiently reliable supply of potable water. It was necessary to consider the costs associated with alternative methods of constructing this scheme. The principal components of the scheme are the catchment area, used to collect the rainwater, and the reservoir, used to store it.

The rain which falls on the catchment must flow by gravity into the reservoir. Site topography determines whether or not water can be withdrawn from storage by gravity. For relatively level terrain the reservoir can supply water by gravity only by being above ground level: this requires the catchment area also to be above ground. If the catchment is at ground level, on level terrain, the reservoir has to be excavated and water can then only be withdrawn by pumping.

SITE CONDITIONS

Manda Island is accessible by sea and by air. The nearest point on the mainland, Mokowe (which has a large jetty at which ships of 200 tons can berth) is within 3 km. Transportation constraints made it almost impossible to consider using heavy construction equipment for water supply facilities on the island.

The area allocated by the government for an agricultural settlement on Manda Island is a flat area with slopes of about 1.5 per cent. A shallow layer of topsoil (averaging 1 metre in depth) overlies the coral of which the island is composed. The coral can be excavated slowly by hand or more efficiently with pneumatic hammers. Gravel and sand for use in structural concrete are not available on the island. Large quantities of fine coral sand are available on ocean beaches within 2 km of the rain harvesting site but the salt content in this sand makes it unsuitable for good concrete.

Land on which the water supply was to be built was assumed to cost $180 per hectare or approximately $0.11 per square metre.

The difficulties of transporting and operating construction equipment and the availability of large numbers of unskilled labourers (earning approximately $1.50/day) made it desirable to use unsophisticated construction techniques. Skilled labour was obtained from Mombasa at a cost (including accommodation) of approximately $5.00/day.

PROTOTYPE SCHEME AT WASINI ISLAND

In Kenya, Grover (1969) had the opportunity to construct a prototype rain harvesting scheme on Wasini Island, approximately 80 km south of Mombasa (near the Tanzanian border) at a site very similar to Manda Island.

Prior to the construction of this prototype scheme, drinking water was taken to the village in dugout canoes from wells on the mainland, about 2 km away. Water imported in this manner augmented the limited amount of freshwater collected in several small rain harvesting schemes constructed in earlier times. These consisted of excavated and plastered tanks (approximately 20,000 litres each) with the excavated material piled alongside and plastered to form a small catchment area (approximately 9 square metres). These individual schemes were in poor repair, leaked badly and could not provide the water requirements of the population in the dry season.

A large catchment area and reservoir was then proposed as a central water supply for the village on Wasini Island. Using principles similar to those outlined in chapter 4, an excavated reservoir was planned with a catchment at approximately ground level to supply the reservoir. Soil from the excavation was to be used to eliminate depressions and improve the slope over the catchment area.

Preliminary designs and cost estimates for the Wasini prototype indicated that the most economical means of increasing the runoff from the previous material on the catchment was to seal the surface with sprayed asphalt. The excavated coral had to be covered first with enough sand from the nearby beach to provide a reasonably smooth surface for runoff.

Simple tests are arranged at the Water Development Department headquarters in Mombasa, Kenya, to determine the best means to seal the sand surface. A membrane formed by spraying various asphaltic materials directly on sand was impervious but was unsatisfactory as a surface because it was too easily damaged. It could be peeled off the sand, or would crumble when walked upon.

The sand which was the base for the impervious surface had to be strengthened. Tests indicated that this could be achieved by mixing modest quantities of cement in the top few cms of the sand and moistening the mixture to form a type of soil-cement. When the cement set, the sand was sufficiently firm and cohesive to support the workers who would have to apply the waterproof surface.

Various asphaltic materials were considered for making the surface impervious. Tests indicated that a combination of two locally available products which could be applied without heating worked best. The first spray applied, the seal coat, consisted of a medium-curing cutback asphalt (MCO) which is asphalt cement in solution with kerosene (about 1 kg asphalt per litre). This solution penetrated the top layer of the stabilized sand before the solvent evaporated, leaving behind a light deposit of asphalt.

The second and final application was spray consisting of emulsified asphalt (*Colas* trade name) which is asphalt cement in an emulsion with water (about 1.1 kg asphalt per litre). The second spray bonded well to the seal coat and when the water evaporated, after a period of several days, a good waterproof surface resulted.

For the Wasini reservoir it was decided arbitrarily to excavate a volume of about 311 cubic metres (to store 320,000 litres of water) out of the coral. This excavation would then be lined with a single sheet of butyl rubber.

Excavation of the reservoir had already been started by the local people on Wasini Island before the preliminary design was complete. Sides of the reservoir were sloped at about 60° to the horizontal. The coral would allow even a vertical 90° excavation but the supplier of the reservoir liner recommended the 60° slope to prevent the butyl rubber liner from stretching when all edges were anchored at ground level. The reservoir was square in plan, 12 m on each side at ground level and 9 m at the bottom of the 3 m deep hole. Because the surface of the excavation was jagged in many places and might have punctured the butyl liner, the bottom and sides were smoothed by filling holes in the surface with rough plaster made of beach sand and cement. The reservoir was lined with a single sheet of butyl measuring 18 m^2 (9 mm thick). All edges of the sheet were anchored in trenches, which were backfilled with sand, along the reservoir edge.

Spoil from the excavation was sufficient to smooth an adjacent area of about 1,400 square metres. This determined the extent of the catchment. Beach sand was used to cover the coral spoil and eliminate minor depressions in the surface. The catchment was first dampened with sea water and compacted with a small vibratory roller (used normally for highway surface repairs) before being covered with a 8 cm layer of soil cement, consisting of beach sand mixed with cement 6 per cent by weight. This proportion of cement is in accordance with the 3 to 10 per cent range recommended for road construction using granular soils (Yoder, 1959).

The seal coat of MCO was sprayed on the stabilized sand at an application rate of about 1.4 square metres per litre. The asphalt solution was pumped from 200 litre drums using a hand sprayer on a four wheel carriage. Several days later the second application, consisting of Colas, was sprayed at an application rate of about 0.7 metres per litre.

A hand pump was installed to withdraw water from the reservoir. A simple roof was installed over the reservoir by suspending a second sheet of butyl rubber on ropes anchored at the reservoir edges to provide a grid support pattern for the butyl sheet. Small holes, centred in each grid of the roof cover, were used to allow rainwater to enter the reservoir without being trapped on the roof. Unfortunately, this part of the Wasini prototype structure failed. The ropes sagged and the holes intended to pass rainwater were offset from the centre of each grid section. When heavy rains fell the water which could not enter the reservoir through the offset holes remained on the cover. This caused the ropes to stretch further and eventually the sagging butyl sheet ripped in several places.

It is interesting to note the simplicity of the construction methods which had to be used at Wasini Island. All material had to be taken to the island in small boats: the 340 kg vibratory roller was very difficult to transport.

Experience gained in the construction at Wasini Island indicated that this prototype was a practical solution to community water supply problems in similar situations along the African coast.

RESERVOIR FOR MANDA ISLAND SCHEME

As the site at Manda Island was similar to that at Wasini Island, a community water supply similar to the Wasini prototype was proposed, with a catchment at ground level supplying runoff to an excavated reservoir. Because the coral in which the reservoir was to be excavated was pervious the reservoir would have to be lined.

Reservoir linings of plastered masonry or concrete were considered but rejected on the basis of cost. Flexible membranes of various types would be cheaper. On the basis of experience elsewhere vinyl and polythene were also rejected because of their relatively short service life; they are simply not tough enough. Butyl rubber, however, has been tested as a water barrier in many applications and has been found to have excellent characteristics, being waterproof and durable over long periods of time, even when exposed to sunlight (Lauritzen, 1967).

A single prefabricated sheet of butyl rubber was proposed as the liner for the Manda reservoir. The excavated surface of the reservoir was smoothed as required to eliminate jagged edges which might cause tearing of the liner. Whenever tears or punctures occurred during construction, patches were welded to the sheet to keep it watertight.

To minimize land requirements and potential surface evaporation the reservoir was made as deep as possible. The cost of excavation increases with the depth however, and simple hand pumps which would be used to withdraw water from the reservoir are limited in practice to a suction head of about 5 metres. The reservoir depth selected in this circumstance therefore was 5 metres.

A novel method of covering the reservoir was proposed. To avoid the costs of a rigid structure supporting a roof it was proposed to cover the reservoir with a single sheet of butyl rubber attached to flat pieces of polystyrene which would float on the reservoir surface (with a specific gravity of 1.25 the butyl would sink unless buoyed up). The floating cover would be impermeable and would rise and fall with the level of the reservoir.

There were no holes in this top cover. This would reduce the risk of pollution but would waste all rainwater falling on the reservoir surface: it would eventually evaporate, since potential evaporation is roughly three times the annual average rainfall in this area.

The cost of providing a cover for the reservoir cannot be justified on the basis of evaporation prevention alone, because sufficient extra catchment area to replace the runoff lost annually through evaporation could be constructed more cheaply than the reservoir cover. The floating cover, however, had several other advantages:

(a) prevention of algae growth in the reservoir by elimination of sunlight on the water surface;
(b) prevention of airborne pollution from seabirds; and
(c) visual indication of water level (psychological assistance for water conservation in drought periods).

On the basis of the costs of labour and materials expected at Manda Island, the approximate cost of excavating, lining and covering the reservoir in the manner proposed was $7.14 per cubic metre. This cost included an allowance for placing the excavated material on the catchment. In the range of storage volumes required for the 10,900 lpd supply at Manda Island this unit cost would be reasonably constant.

With butyl rubber sheets for lining and covering the reservoir it was assumed that water losses from the reservoir due to evaporation and seepage would be nil.

THE CATCHMENT FOR THE MANDA ISLAND SCHEME

The catchment was at ground level, sloping towards the excavated reservoir. For Manda Island, the following methods (table 6-4) were considered for preliminary cost estimates:

Table 6-4 Cost of catchment surfaces

Catchment surface	$ Cost/m^2
Corrugated asbestos cement sheets	5.40
Precast concrete paving stones	4.80
Concrete slab (cast in place, 8 cm thick)	4.20
Butyl sheet (0.74 mm thick)	3.11
Sprayed asphalt on stabilized base	1.61

These preliminary estimates and the experience on the Wasini Island prototype indicated that the last alternative, i.e. sprayed asphalt on a stabilized base, was practicable and the most economical. The method used at Wasini Island for preparing the area was proposed for Manda Island. Excavated material from the reservoir was placed in depressions in the area of the catchment and used to increase its slope. The area was compacted by a vibratory roller before the top 8 cm layer was mixed with cement (approximately 6 per cent by weight) to stabilize the base. This mixing was done accurately, batch-mixed just like plaster. This method produced excellent results but was time consuming and labour intensive. It should have been possible to achieve equally satisfactory results at Manda Island by spreading measured quantities of dry cement on the sand surface and mixing it with rakes and hand tools. The high cost of using machinery and the low cost of labour would preclude the use of any mechanical plant except for the roller.

A drawback to using an asphalt-covered catchment was that the runoff was frequently coloured as a result of deterioration of the asphalt. The coloured water was usually odourless and tasteless and has been consumed by people (e.g. at Kilifi on the Kenyan coast) with no known ill-effects but not enough is known to state categorically that it causes no problems. Asphalt is classified as a non-toxic material (Gleason *et al.,* 1957).

The breakdown of asphalt into the water-soluble degradation products which discolour runoff is caused by a combination of light, heat and oxygen. This degradation can be reduced by protecting the surface of the asphalt from sunlight: a layer of stone chippings frequently performs this function on roofs waterproofed by asphalt. Recent research indicates that sprayed asphalt catchments can be protected by a spray incorporating flaked aluminium (Frasier *et al.,* 1968). For Manda Island it was proposed to spray a cover of this material on the final asphalt surface to prevent degradation of the asphalt.

Using the construction methods tested at Wasini Island and improved upon by spraying a protective layer on the asphalt, the estimated cost of constructing the impermeable catchment area was $1.61 per square metre.

The quantity of runoff from a catchment depends on how much

precipitation is lost on the catchment through depression storage, infiltration and evaporation. The total runoff from a given area can be assumed to be equal to the total volume of precipitation falling on the catchment, reduced by a runoff coefficient. This can be expressed mathematically as:

$$R = K\,P\,A \qquad (1)$$

where: R = total runoff in period (volume)
K = runoff coefficient
P = total precipitation on catchment in period
A = area of catchment

Equation (1) can be obtained by integrating over time the standard expression for rate of runoff (rational method).

There is plenty of discussion in the literature of urban hydrology concerning values for the runoff coefficient for use in determining peak flow rates. It is obvious, however, that its value will vary for a given catchment depending on the intensity of precipitation. (Runoff for a light drizzle can approach zero.) What is required for a rainwater harvesting scheme is a constant 'R' which can be applied to all precipitation occurring within a period. While obtaining such a coefficient may be desirable it is also extremely difficult. It will depend on a number of factors which apply to a specific site, including:

(a) precipitation patterns (variations in intensity);
(b) slope of catchment;
(c) smoothness of catchment; and
(d) permeability of catchment.

Since a rainwater harvesting scheme is constructed to maximize the runoff from the catchment, the runoff coefficient should be relatively high. Researchers in Arizona have measured runoff from asphalt pavements similar to those proposed for Manda Island and have found runoff ranging from 96 to 101 per cent of measured precipitation over one year (Myers *et al.,* 1967). These experimental catchments were in areas with annual rainfall averaging only 20 cm. The catchments were 233 square metres in area and had slopes of 5 per cent. The explanation for runoff being more than 100 per cent of measured precipitation is that the rain gauges under-registered the actual precipitation at the site.

No measurements of runoff were made at the site of the catchment built for the Wasini Island prototype. In the absence of better information, it has been assumed that the catchments proposed for Manda Island (with annual precipitation averaging about 91 cm and catchment slopes of 1.5 to 2 per cent would have a runoff coefficient of 0.90, and this is a conservative estimate of the coefficient.

DESIGN OF THE SCHEME

In a conventional water supply using a surface water source the catchment area is defined by the regional topography. The design problem in such a case is basically to determine the necessary reservoir volume to meet the estimated water demand for expected runoff conditions. Reservoir volume is the only design variable.

In a rainwater harvesting scheme there are two design variables: catchment area and storage volume. Many combinations of sizes of catchments and reservoirs can supply the estimated demand. Since multiple solutions are technically possible, selection of the optimum solution requires determination of the least cost solution. A complication arises, however, since not all of the appropriate combinations of catchment area and reservoir volume would provide a water supply having the same reliability.

RELIABILITY OF WATER SUPPLY

The future rainfall on which a rainwater harvesting scheme depends can only be estimated. Normally this is done by referring to records of historic precipitation. Frequency curves of historic precipitation indicate that it is virtually impossible to ensure the provision of 100 per cent of the demand for water over 100 per cent of the time from a water supply based on runoff unless an infinite amount of storage is provided. The designer of a rainwater harvesting scheme needs to be aware of the reliability of the supply which it is possible to provide by the various alternative combinations of catchment area and reservoir volume.

It is apparent that a precipitation harvesting scheme can be made more reliable by increasing its catchment and/or storage. These improvements cost money. To select the appropriate degree of reliability the designer should somehow be able to evaluate this reliability in money terms. The scheme would then be increased in size until the incremental cost of increasing the size of the scheme exceeds the incremental benefit of increased reliability of supply.

Unfortunately it is seldom easy to measure the benefits of increased reliability of supply. In the case of Manda Island, however, there was a means of evaluating this aspect of the problem.

The per capita water consumption of 23 lpd estimated for the Manda Island settlement was relatively low. Any reduction in this supply would create major difficulties for the population. Nevertheless the people could survive periods of reduced supply, by one or more of the following means:

(a) voluntary water rationing;

(b) using alternative sources for water, such as importing water; and/or
(c) leaving the area temporarily and moving to an area with an adequate water supply (e.g. Lamu).

If the proposed rainwater harvesting scheme on Manda Island ran dry the people would probably restrict their consumption and import water by boat. It has been assumed that imported supplies would have to meet at least 60 per cent of the estimated requirements. For total failure of the Manda Island supply this would mean importing a total of about 16,000 lpd. Emergency supply could be provided from Lamu in drought periods. Alternatively the water could be imported from Malindi.

Rather than select an arbitrary degree of reliability which each water supply must provide (e.g. at least 70 per cent of the supply for 100 per cent of the time) the reliability has been treated as a cost function. For each day, when the full supply cannot be provided by rainwater harvesting, it was assumed that supplementary supplies to provide at least 60 per cent of the demand are obtained at a daily cost of $70. This assumption over-simplifies the actual situation slightly, since a partial supply of water from rainwater harvesting schemes reduces the amount of water which has to be imported. Setting up the temporary measures to augment the water supply for Manda Island would probably involve constant daily costs, even if smaller quantities of water were delivered.

SELECTION OF OPTIMUM CATCHMENT AREA AND RESERVOIR VOLUME

The rainwater harvesting scheme for Manda Island was designed on the assumption that precipitation records for Lamu over fifty-eight consecutive years (1911-1968) are indicative of the precipitation which can be expected in future on Manda Island.

Since both catchment area and reservoir volume are design variables, calculations were required to determine the dimensions of the optimum scheme. The method of calculation for each model was straightforward. First a catchment area was selected. For that area several trial calculations were made with reservoirs of different capacity. For each reservoir capacity the reliability of the supply was estimated by determining the shortages which would have occurred with the precipitation of the period of record. Costs associated with constructing the schemes and providing alternative supplies of water during periods of shortage were considered to determine the optimum scheme.

For each of the four model catchment areas (from 5,500 square metres to

9,300 square metres) at least four reservoir models were also used to determine what shortages would occur over the fifty-eight years or 696 months of record. Naturally the largest reservoirs were able to provide the more reliable supplies of water, but in none of the twenty trials was water supplied 100 per cent of the time. A frequency curve was plotted to relate the shortages to the reservoir volume for each catchment area. The results (figure 6-1) show a family of curves which fit the data quite well.

Following compilation of the frequency curves of shortages associated with various combinations of catchment area and reservoir volume, it was fairly easy to determine the least cost solution for the rainwater harvesting scheme. The three principal costs which varied for each combination were:

(a) construction costs for the catchment area;
(b) construction costs for the reservoir; and
(c) costs of importing water during shortages caused by drought.

For each of the trial values of catchment area the minimum total annual cost is underlined. Larger reservoirs cause higher total costs because of greater construction costs and smaller reservoirs cause higher total costs because of increasing shortages in supply.

The data indicate that the minimum total annual cost occurs with a catchment area of 9,300 square metres, and a reservoir volume of 2,200 cubic metres. A rainwater harvesting scheme of these dimensions could have supplied 10,900 lpd with shortages occurring only 1.03 per cent of the time (about seven months in total over the period 1911-1968). All other combinations of catchment and reservoir resulted in higher annual costs.

The relative ease of determining the optimum size of the principal components of the precipitation harvesting scheme by using the variable scale mass curve technique should be emphasized. All calculations, including plotting of data and repetitive graphical analyses, could have been carried out for the Manda Island situation in approximately one week of work by a single person (once cost data were available on the basis of preliminary designs). The results clearly indicate the optimum solution for the given parameters, but basic assumptions could be varied and new solutions obtained without much extra work. For example, sensitivity analyses could easily be carried out to determine how various interest rates affect the determination of the optimum project size.

It would be possible to define the size of the catchment area and/or reservoir volume for Manda Island more precisely by repeating the analysis for smaller increments of size and by plotting the curve of total annual costs to determine the minimum value. For the meteorological data and preliminary cost estimates available, however, further refinements to the calculations were not warranted in this case.

Figure 6-1 Frequency curve of water supply for various catchment areas

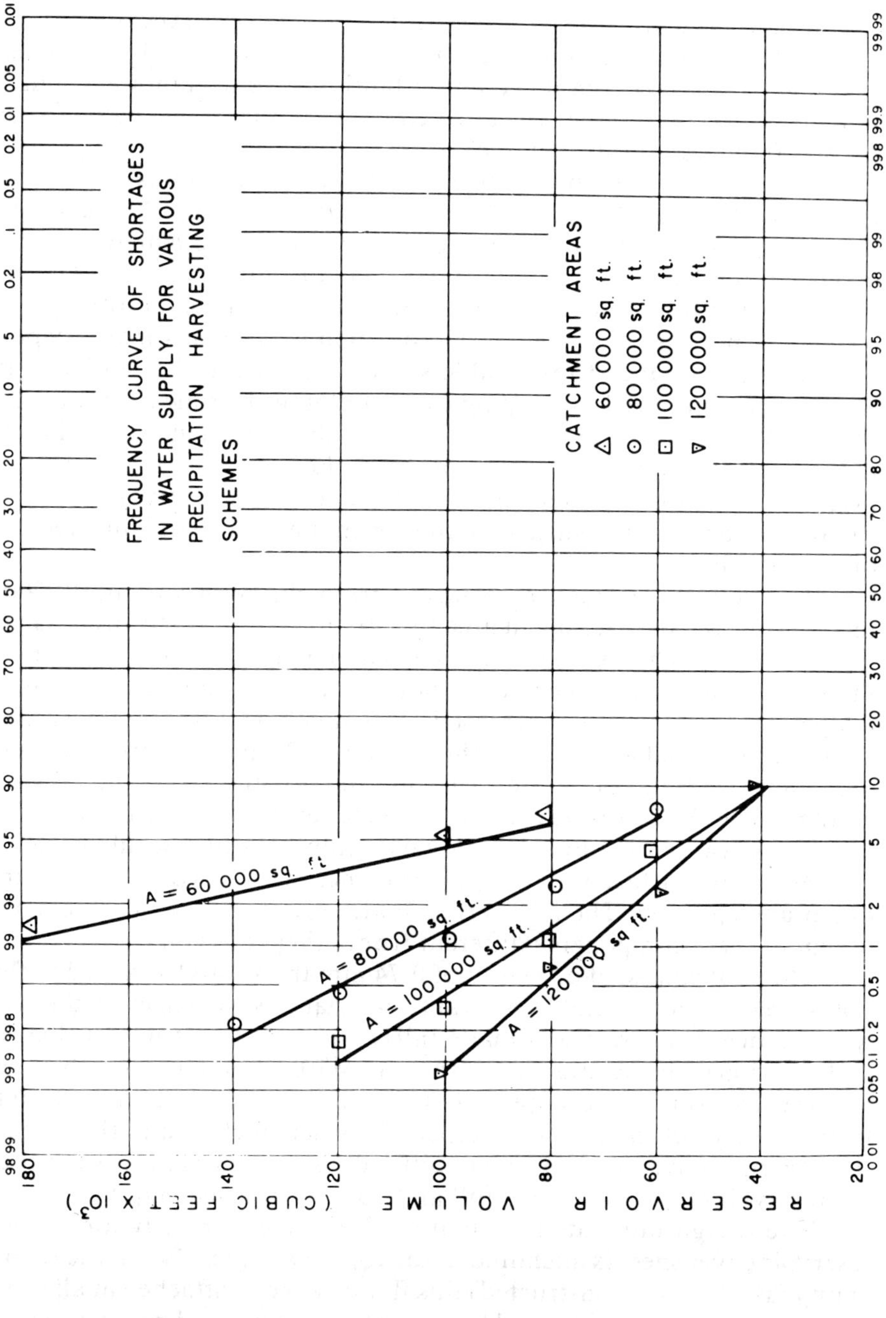

CONSTRUCTION DETAILS

Because the principal elements of the rainwater harvesting scheme were the catchment and reservoir, once the size of these elements had been decided, the final design was completed and attention paid to minor elements in the scheme. The details of the typical scheme for Manda Island, having a capacity of 10,000 lpd, are illustrated in figure 6-2.

The required reservoir volume of 2,200 cubic metres was obtained with an excavation 5 metres deep, 24 × 24 m at the surface and 20 × 20 m at the bottom (giving side slopes of about 30° to the vertical). The catchment area was arranged to drain runoff into one edge of this reservoir. The specific configuration of the catchment area depended on site conditions. A regular shape was not essential. The optimum catchment avoids major depressions and takes advantage of favourable slopes adjacent to the reservoir. Runoff losses were minimized by keeping to a minimum the time for runoff to enter the reservoir. The best theoretical layout for the catchment would therefore be a circular area with the reservoir in the middle. Since the roof arrangement favours collection of water on only the edge of the reservoir, the layout in figure 6-2 was a compromise with this optimum, using straight edges for ease of construction.

The volume of material removed from the reservoir site (2,200 cubic metres) covered the catchment area (9,300 square metres) to an average depth of about 25 cm. This excavated spoil was used mainly to eliminate depressions in the catchment area. The natural slope at the site, which was assumed to be 1.5 per cent could not be increased to as much as 2 per cent even if all the excavated material was used for this purpose. The material from the reservoir thus improved the runoff coefficient, mainly by eliminating depression storage: its effect on reservoir slope concerns the velocity of runoff, which influences evaporation losses from the catchment, which would be negligible.

As in any surface water-supply scheme, the reservoir required an inlet, outlet and spillway. The innovative floating roof for this scheme makes the design of these components different from other reservoirs.

The roof consisted of a sheet of 0.74 mm thick butyl attached to floats of polystyrene or any suitable buoyant material. As the edges of the reservoir were vertical the cover was made fairly rigid so that it would float on the surface without being attached at the top. With a butyl liner for the reservoir, the reservoir sides were sloped to avoid excessive tension on the butyl sheet from the sides of the reservoir, especially when anchored at the top edge. A floating rigid cover would run into difficulties with the varying surface area of the reservoir (24 × 24 m when full and 20 × 20 m when empty).

The design illustrated in figure 6-2 envisages the situation where the reservoir cover sheet is anchored at the top on all sides. When the reservoir is empty (as when it is constructed) this floating sheet, attached at all top edges, lies along the sloping sides and bottom of the reservoir. From one edge to the

Figure 6-2 Design for an experimental catchment and reservoir for Manda Island

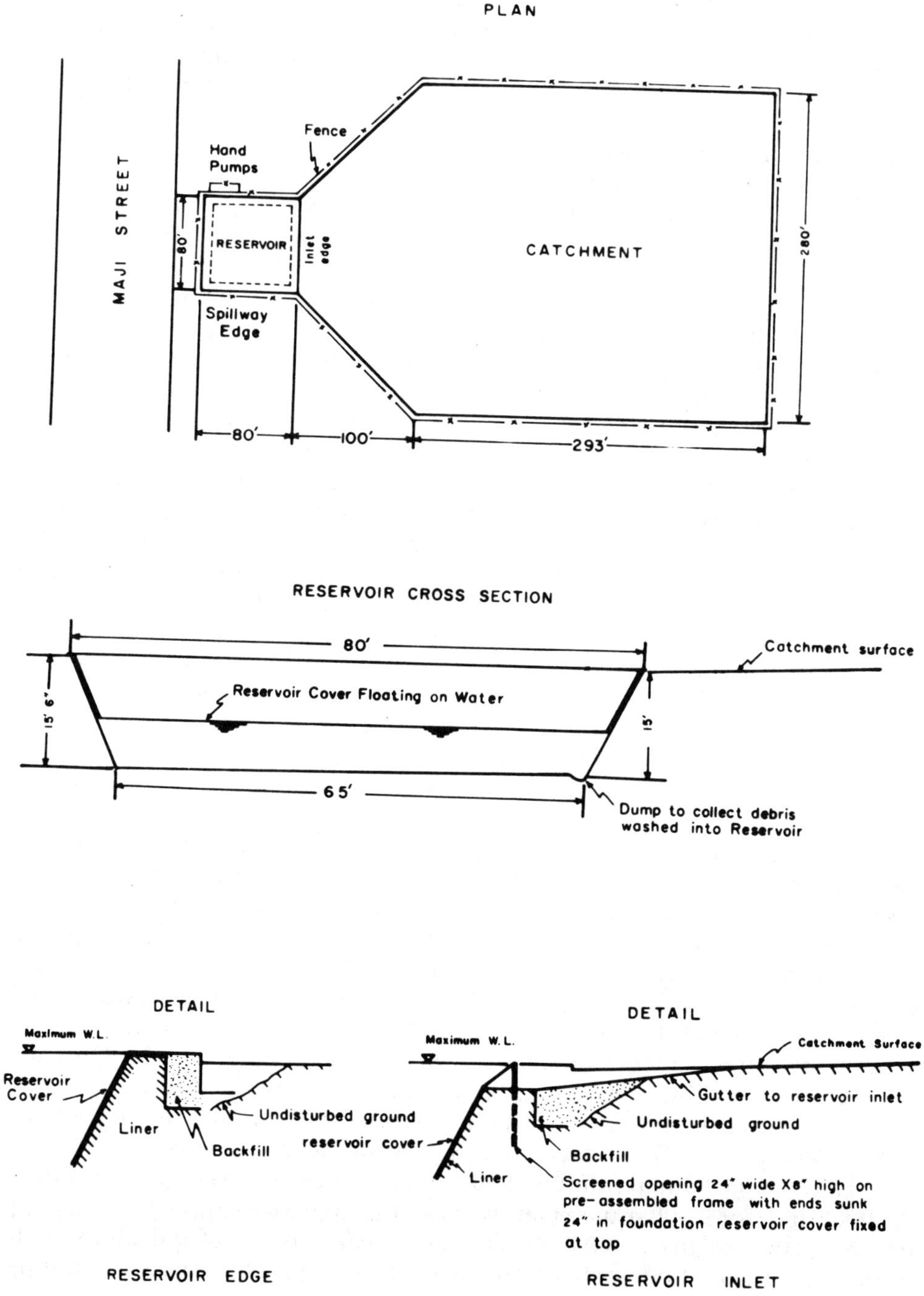

opposite one the length of the sheet would be 30 m when the reservoir is empty. As the reservoir fills the floating cover would rise until the distance between opposite edges becomes only 24 m, which is the width of the reservoir. The slack in the floating cover when the reservoir is full would result in a crinkled sheet, possibly not too pleasing aesthetically, but functionally adequate. If a good reason were found for smoothing this top sheet the slack could be taken in as the water level rises (and conversely released when the level drops) but it seems difficult to justify such a chore.

All edges of the reservoir liner, and two edges of the top sheet, would be fastened by placing the edges in trenches and backfilling to anchor the sheet in place. Arrangements for the two other edges of the reservoir cover were somewhat different.

The inlet edge of the reservoir, adjoining the catchment, requires openings to allow runoff to enter storage. Specially constructed inlets were mounted, and equally spaced on the edge of the reservoir adjoining the catchment. Each inlet had a pre-assembled rustproof frame, with screening to prevent trash from entering the reservoir. A sheet of butyl hanging on the inside of this screen acted as a flap valve, preventing any evaporation loss and also preventing the entry of insects into the reservoir. The catchment was contoured so as to direct runoff towards each inlet. Ten such inlets (61 cm long by 20 cm high) would allow 0.5 m^3 per second to enter the reservoir, assuming that each inlet acts as a broadcrested weir with a depth of flow of 15 cm and a discharge coefficient of 2.5. This runoff was expected for the catchment area of 9,300 square metres, with a rainfall intensity of some 19 cm per hour, a torrential downpour which would be expected only infrequently. The reservoir cover was 20 cm above the inlet level and 5 cm above the general catchment level, giving additional freeboard to prevent storm runoff overflowing into the top of the reservoir. As the reservoir cover was floating on the water surface, however, no structural damage would result in such a case. The only loss would be the volume of water which overflowed.

If the reservoir had no spillway, the runoff from the catchment would back up, flood the spillway and overflow on to adjacent areas (and possibly the reservoir cover) when the reservoir is full. Since erosion damage would be possible in the vicinity, and water lying on the catchment might be subject to pollution prior to slowly entering the reservoir, a spillway was therefore considered essential. For simplicity the same arrangements used at the inlet portals were used at the outlet edge for spillway. Screened and preassembled frames prevented rodents or debris from entering the reservoir. On each spillway portal the butyl sheet acting as a flap valve was placed on the outside of the screen. A small overflow channel was located parallel to the spillway edge to lead the overflow away from the site and to prevent erosion.

The average annual rainfall of 900 mm produced an average daily inflow of 20,800 lpd (for 9,300 square metres of catchment with a runoff coefficient of 0.9). Since the design capacity of each scheme was only 10,800 lpd, almost half of the runoff would be wasted over the spillway. This waste water would be reduced by increased consumption (extra water for laundry, etc.) at times when the reservoir is full, but strenuous efforts would be required to remind the settlers that this excess water could not be provided regularly.

The function of the entire scheme was to provide water for people. As the reservoir was below ground level, water would have to be removed by pumping. The settlers on Manda Island generally went to collect their water twice daily, morning and evening. Assuming that total daily water requirements of 10,800 lpd are to be withdrawn from a given water supply in two periods of two hours each, the pumping rate was about 2,700 litres per hour or 45 litres per minute. Two pumps were installed, to reduce queueing and also to provide for occasional maintenance to either pump. Simple and robust hand-pumps were mounted near the corner of the reservoir, with plastic suction pipes laid along the sloping reservoir wall to the reservoir bottom. The bottom of the reservoir sloped gradually from the inlet edge so that its depth at the pumps was some 15 cm greater than on the inlet edge, ensuring that all water could be removed from the deep end of the reservoir in times of drought. To facilitate the settling of larger particles liable to be washed into the reservoirs, a sump or silt trap was arranged on the inlet edge as indicated in figure 6-2. This facilitated reservoir cleaning by concentrating larger particles for removal.

The entire catchment and reservoir area was fenced in order to:

(a) prevent damage to facilities by domestic animals or children;
(b) minimize pollution on the catchment; and
(c) conserve the water supply by preventing unauthorized withdrawals from the reservoir.

The methods used to construct the facilities were similar to those in the prototype built at Wasini Island.

WATER QUALITY

The purest water found in nature—rainwater—is stored in a rainwater harvesting scheme within minutes of its arrival on the surface of the earth. The catchment surface was constructed to produce maximum runoff with minimum change in water quality, and the aluminium spray mentioned earlier was successful in eliminating degradation of the asphalt, thus making the water in storage free of discolouration from this source. When such discolouration existed on previous asphalt catchments the water was nevertheless odourless and tasteless. Discolouration was caused by materials comprising less than 10 mg per litre (Myers *et al.,* 1967). Colour removal is difficult, requiring ionic exchange resin columns.

Two potentail sources of pollution were considered to be:

(a) wind-blown particles (sand, etc.) which could be washed into the reservoir from the catchment; and
(b) depositions from birds (mainly seabirds).

The constructed fence around the catchment prevented it from being made dirty in any other manner. Screens on the inlet portals kept out most matter but some sediment could be expected to enter storage. The reservoir, however, acted as a very quiescent settling basin and any material washed into it soon settled to the bottom. With the pump suction located above the reservoir floor the water withdrawn from storage contained little suspended solids.

The reservoir liner and cover made of butyl was unaffected by the water stored. It appears true also that the water stored is not affected by this butyl.

Algae, which can cause aesthetic and odour problems in reservoirs, require sunlight for their photosynthesis process (Palmer, 1962). The reservoir cover does not allow water in storage to be exposed to light, and runoff moves quickly across the catchment in rainy periods so algae are not likely to exist in the harvesting scheme.

To maintain water of high quality in this scheme required simple attention to prevention of pollution. All people and animals were kept completely away from the catchment and reservoir, by fencing and by administrative enforcement. Regular maintenance activities were performed by trained personnel so that the water supplied by the scheme remains wholesome and palatable.

There was no treatment required for mineral or organic matter in the store of water collected by the scheme. Since there was no reason to expect any poisonous substances or disease-producing organisms to enter the water of this simple scheme, it was assumed to be hygienically safe. In these circumstances no provision has been proposed for any kind of water treatment at Manda Island.

If there were any reason to suspect bacteriological contamination of the water supply (by known entry of fecal pollution into the reservoir, for example), the immediate solution to avert a public health problem would be simply to stop using the particular scheme which was affected. Two of the three schemes should be unaffected since the pollution would probably be only local.

The arrangement proposed for covering the reservoir has a possible disadvantage in that the black butyl sheet, in direct contact with the water except where the polystyrene floats are attached, would absorb and transmit heat from sunlight. The thermal conductivity of butyl has not been investigated but it is reasonable to expect that the water stored in the reservoir will be somewhat warmer than the air in the area. Effective prevention of evaporation from the reservoirs by the floating cover would prevent cooling of the body of water through evaporation from the surface. If the temperature of the water were sufficiently objectionable the consumers might cool it through controlled evaporation, possibly by using canvas water bags for water transport or storage.

MAINTENANCE PROCEDURES

Minimal maintenance is required in the scheme. The catchment area requires occasional sealing of cracks and a resealing of the entire surface would probably be desirable, perhaps on an annual basis. This is done by spraying with emulsified asphalt or, if the final spray incorporating flaked aluminium succeeds in preventing degradation of the asphalt, this top layer of aluminium is simply resprayed. In either case, the work involved is done by one man in one day using spraying equipment which was obtained during construction.

The fence has to be kept in a state of good repair and the hand pumps require occasional servicing. Such simple jobs are completed as and when required.

The reservoir requires inspection and removal of sediment periodically. Most of this sediment accumulates in the small sump at the intake edge of the reservoir. This material is removed most easily when the reservoir is empty, and involves temporarily removing the floating butyl cover. At that time a visual inspection is made of the reservoir liner, pump installation, reservoir cover and float. The reservoir is designed, however, to be seldom empty. As the settlement will have a total of three schemes, and as each would use about half the normal available runoff, it would be possible to empty one reservoir by pumping before the rainy season so that it could be cleaned out.

The butyl liner and cover of the reservoir should remain perfectly watertight after installation. The cover is the more vulnerable element. If the cover or the liner should tear or develop a hole, the necessary repair can be made fairly simply by patching. A supply of patches, tape and adhesive for repairing the butyl sheets was left on the site following construction.

CHAPTER 7

Rain and Stormwater Harvesting by Contour Terracing

TERRACING

TERRACING is simply an application of rain harvesting but on a large scale. Low walls are built along the contours of the slope from stones and rock fragments. Their uphill sides are filled with soil collected from the slope between the terraces. The surface thus becomes bare, and ensures the complete runoff of rainwater. The length of slope between terraces is relative to the horizontal width of arable strips in between, and this in turn depends on the amount of precipitation and on the water demand of the crops used.

Traditionally slopes were cleared to prevent the retention of water by vegetation and the catchment was divided into smaller units of 1 to 3 ha by constructing simple channels. The purpose of this operation was to accelerate the runoff process and to decrease evaporation loss. The units were small enough to prevent erosion on the slopes and to prevent the development of high, uncontrollable discharges in the channels. The water was either stored in the soil of the cultivated land, or sometimes artificial storage facilities were constructed, such as open, lined tanks or closed cisterns. In some cases the storage capacity was increased to very large volumes.

In many rural areas storms of high intensity produce flash floods in torrential watercourses and the water then runs down without utilization. The possibility of storing this water in the form of soil moisture and/or groundwater over large areas by diverting it was recognized at early stages of socio-economic development throughout the world. The diverting structures and the distribution systems devised to spread this excess water over large areas of the flood-plains are simply canals and low rock-fill dykes. This topic will be developed in chapter 9.

Examples of contour terracing are known from many parts of the world. The use of terracing in China will be taken up in chapter 11, but in this present chapter it will be worthwhile to look in detail at the contour terrace systems found in southwestern USA and in Central America.

Contour terraces known as *metlepantli* or *bordos* are found on gentle and medium grade slopes of piedmont areas throughout the central and southern highlands of Mexico. Although the extent of their use appears to have diminished since pre-Spanish times, contour terraces are still the most extensive form of water harvesting in Mexico (Spores, 1969).

A growing number of studies focusing in detail on contemporary terraces include Patrick's work in Tlaxcala (1977); West's work in the Valley of Mexico and Hidalgo (1970); Johnson's work in the Mezquital Valley (1977); Sanders' and Charlton's work on the Teotihuacan Valley (1970); and Wilken's review of traditional forms of slope management (1976).

As described in chapter 2 the ancient contour terrace systems consist of placing long, low barriers at even intervals perpendicular to slope gradients. These barriers can be rows of stones, logs, earth embankments or hedges of agave (*maguey*). Often these materials are found in combination, particularly when economic plants are used to reinforce earthen embankments or stone walls (figures 7-1, 7-2 and 7-3).

The main aim of these structures is to interrupt and retain slope runoff and alluvium. Over time, alluvium builds up behind the barriers thereby increasing infiltration and enhancing the soil moisture storage capacity of the fields. As shown in figure 7-4, contour terraces may also include drainage ditches (*zanjas*) whose function is to trap and store any excess runoff which overtops the upslope embankment.

Most contour terraces are located on slopes of less than 25 per cent. Unlike other forms of terracing, the elongated, sloping fields do not change slope gradient to any significant degree, but instead basically conform to the existing physical characteristics of the hillside (figure 7-5).

The width of fields varies with slope gradient. Gentle slopes tend to have wider fields and terraces on steep slopes are spaced more closely. One author (Wilken, 1976) recommends the following spacing:

Gradient	*Field Width*
gentle slope (1 to 5%)	5.0 to 6.5 m
moderate (5 to 10%)	4 to 5 m
steep (10 to 25%)	2.5 m

These intervals are somewhat broader than those observed in the field. Characteristically, the ratio between two basic crops planted in *bordos* (maize in the field and maguey on the embankment) varies with gradient. Steeper slopes tend to have a higher proportion of maguey, planted at times at one metre intervals.

One of the most important aspects of contour terracing in Mexico is that it requires very low investments in capital and relatively low investments in labour (figure 7-6). Wilken (1976) observing a government-sponsored

Figure 7-1 Bordo reinforced by useful plants, Mezquital Valley, Mexico

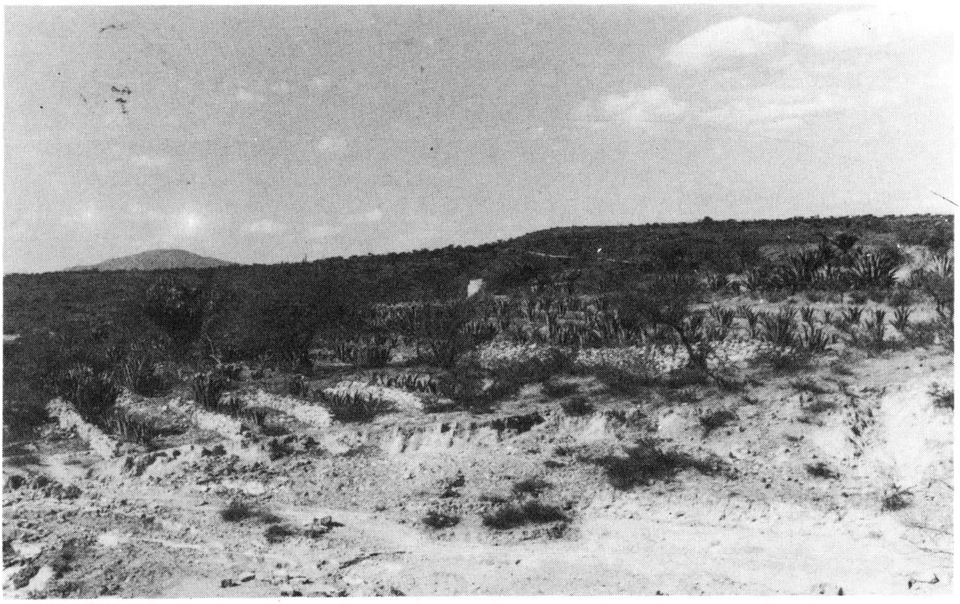

Figure 7-2 Bordos in the Mezquital Valley, Mexico

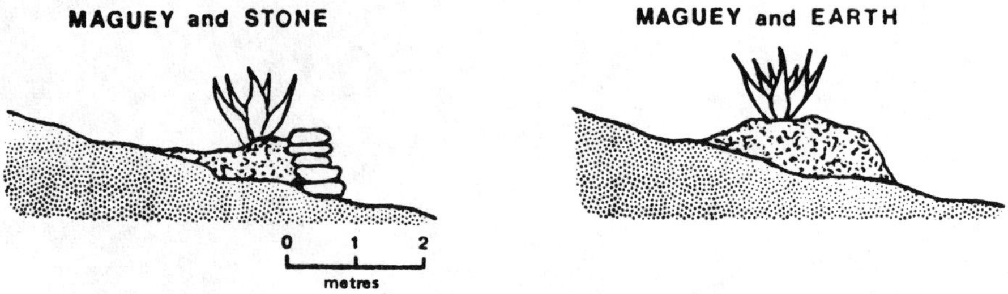

Figure 7-3 Soil barriers built up out of soil, plants and stone. The plant used is maguey which is an agave plant

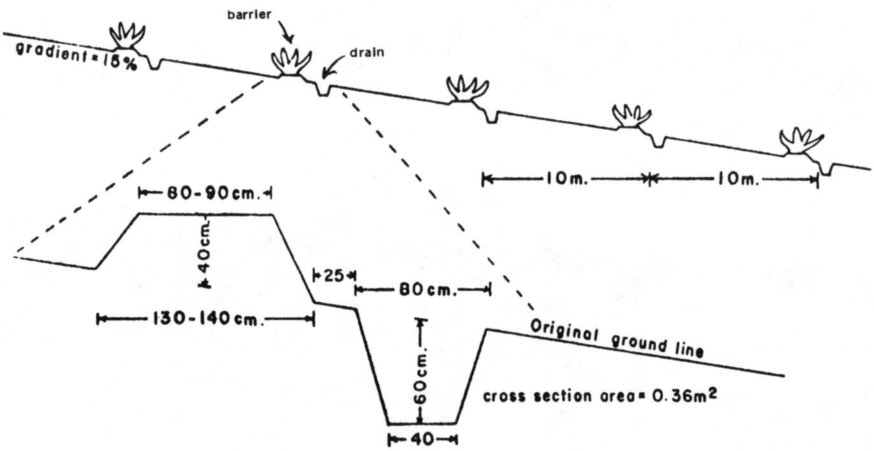

Figure 7-4 Cross section of a contour terrace showing barriers and drains

Figure 7-5 Bordos in the Mezquital Valley, Mexico

Figure 7-6 Bordo construction in the Mezquital Valley, Mexico

contour terrace project in Tlaxcala, Mexico, decribes the process in the following manner:

> On the gently to moderately sloping agricultural lands, *ejiditarios* cut trapezoidal drainage ditches (*zanjas*), 60 cm deep and 80 cm wide at the top sloping to 40 cm at the bottom. *Zanja* lines are laid out along contours then precisely cut with shovels, spades and picks. Dividing strips 50 to 60 cm wide, but only 30 to 40 cm high, are left every few metres to prevent water flow in the *zanjas*. Excavated material is piled immediately upslope in geometrical *bordos* 40 cm high and 80-90 cm wide at the top, sloping to 130-140 cm at the bottom. Material in the *bordos* is not compacted except by incidental foot traffic and thus has a larger volume than the *zanja* from which it came. After *zanja-bordo* construction is complete, young maguey from nurseries are planted along embankments at 3 m intervals.
>
> Workers are paid by the government on the basis of *tareas* or *jornales,* a fair or normal amount of work that can be done in six hours. *Tareas* are based upon working conditions including types of work, terrain, and material. For example, the following are representative *tareas* for excavation and *bordo* construction in Tlaxcala, Mexico:

Material	*Tareas* for *Zanja-Bordo* Construction		
	Tarea in Linear Metres	Approximate Cubic metres (Cross section = 0.36m)	Labour Costs Pesos per Cubic metre (20 pesos/tarea)*
Soft soil (*blando*)	20 - 25	7 - 9	2.85 - 2.25
Moderately compacted (*duro*)	10 - 15	3 1/5 -5 1/2	5.70 - 3.65
Temperate	5 - 10	2 - 3 1/3	10.00 - 5.70
Rock (*rocoso*)	1 - 5	1/2 - 2	40.00 - 10.00

* 1 peso = $U.S. 0.8.

On the basis of these parameters, it can be estimated that the cost in person-days per hectare of contour terracing on moderate slopes (5 to 10 per cent will range from 44 person-days (US $ 70.40) on soft soils to 100 person-days (US $ 160.00) on hard *tepetate.* These costs are low indeed when one considers the efficiency of contour terracing in capturing available runoff, enhancing soil moisture for agricultural fields, and controlling erosion.

For this reason the Mexican government has engaged in a campaign of labour-intensive public works, water and soil conservation programmes that feature the construction of contour terraces.

Contour agriculture lends itself to the use of local labour and is suitable for adoption where the mean annual rainfall is 400 mm or more. In such areas it will definitely help to prevent soil erosion.

The areas where rainfall exceeds 500 mm (especially if 1,000 + mm) a water-harvesting terrace technique has been developed to allow for the conservation of the excess water. This has proven to be especially useful in southern Tanzania and Zambia. Here ditches are dug following contours and the soil is piled on the *downhill* side, creating wide shallow furrows which are so carefully levelled that they will retain excess runoff during the rainy season. This excess slowly seeps into the soil. Seepage in this case is encouraged in clay-pan areas by actually drilling with a soil auger and filling the holes with sand or gravel. With the careful use of spillways to control water levels, this system of furrows can be built up to completely control sheet flow during average rainfall years.

MICROCATCHMENTS

The design of rainwater harvesting systems requires basic data on watershed characteristics, precipitation and streamflow. Maintenance is, of course, necessary and there is a danger of adverse effects of sedimentation and salinization. Nevertheless, there are many potential sites in arid and semi-arid lands suitable for water-harvesting. Of special concern here are microcatchments and sand dune farming. Both of these techniques require some adjustments, but usually not the extensive reworking of the slope as in contour terracing.

The simplest method is to develop a water supply for single plants on steep and generally rocky slopes. This is referred to as the microcatchment method. The unprepared soil surface is generally uneven and thicker soil layers accumulate in depressions. Trees or vines can be planted in these depressions, but their survival depends on the availability of water, which can be provided by rain harvesting even in arid climates. In this method, soil is taken away from a part of the slope uphill to ensure the rapid and complete runoff of rain. A small dyke is formed from rock-fragments, stones and soil below and beside the plant to retain the water running down. The small basin formed in this way is also filled with soil to increase both the amount of arable soil and its storage capacity. The size of the microcatchment to be cleared of rocks to ensure a high rate of runoff. This depends on the water demand of the plant and on the amount of rain and the storage capacity of soil around the roots. An entire plantation of microcatchment can be extended to the whole slope, depending on the required size of each microcatchment, when the distance between plants is determined. Orchards and vineyards were formed in this way many thousands of years ago and the method is still in common use in the Mediterranean region, especially on the Mediterranean islands. The most important plants cultivated in this way are vines and olive trees, but peaches, apricots, figs and almonds are also produced by the microcatchment method.

A modern form of microcatchment farming has now extended to plains areas. Here, the microcatchments are formed by dividing the field into small plots bordered by low dykes. The plants are placed either in one corner of the basin or at its centre. The bottom slopes are developed artificially by earth-moving activity and are made to slope in the direction of the plant.

SAND DUNE FARMING

In the American southwest Indian farmers grow crops on the slopes or base of sand dunes in which rainwater has been trapped in a dense layer of sand located between the loose, dry surface sands and an impervious subsurface of soil or rock (Hack, 1942).

Suitable sites for sand dune farming are often located in the Hopi mesa region of northwestern Arizona. Although the Hopi are best known for this technique, the Zuni of northwestern New Mexico also practised a variant of sand dune farming in the late nineteenth century.

Most commonly, dune fields are located where 15 to 80 cm of sand cover a less-pervious soil or rock subsurface. The loose surface sands trap moisture in a denser sand stratum located between the surface and the substratum. As there is no runoff from sand dunes, all rainwater percolates and is trapped in this fashion in the middle stratum.

Fields are also planted on climbing dunes which have blown against steep mesa walls or against ancient stabilized vegetation-covered dunes. Another preferred location is at the base of a dune where rainwater emerges as seepage.

Common sand dune crops are maize, beans and tree crops. Depending on the field site, it may be necessary to choose crops with deep-rooting systems.

One problem with sand dune farming is erosion. In order to conserve dune moisture for crop use, the natural vegetation must be removed from both the fields and the surrounding dune areas. This encourages dune erosion and exposes tender plants to winds and sand. In less exposed areas, large stones or tin cans are placed around individual plants to protect them from wind and sand. On exposed slopes sand is held down by rows of brush which, in turn, are held in place by rows of stones.

Nineteenth-century accounts (Cushing, 1920) describe the Zuni Indians as sand harvesters as well as water harvesters. Zuni farmers would select a site which could be irrigated by seasonal storm runoff. They would then plant rows of sagebrush windbreaks in barren spots where blown sand and soil would accumulate during windy, dry months thereby creating deposits ready for planting during the rainy season. It is unclear how extensive this practice is today.

In Hopi country, sand dune farming is still important, second only to floodwater farming in extent. A survey in 1937 estimated that 27 per cent of Hopi cultivated land consisted of sand dune fields. One settlement, Hotevilla, is known to be particularly well-suited to sand dune farming.

CHAPTER 8

Rain and Stormwater Harvesting by Silt Traps, Check Dams and Canals

SILT TRAPS

Two important types of stormwater harvesting technologies are now described: (a) silt traps designed to trap both alluvium and runoff; (b) check dams designed to impound water for subsequent agricultural or domestic use.

At times the distinction between silt traps and contour terraces is difficult to make as both are frequently found in association and both function as barriers to trap alluvium and runoff. However, as we shall see, important differences exist between the two types of technologies. Both have been well developed in Latin America and China. In this chapter we will mainly consider examples from Latin America; the Chinese examples will be taken up in chapter 11.

Silt traps (*trincheras*) are built of stone across the beds of intermittent streams, often in narrow valleys, gorges or gullies. As the alluvium deposits build up, level fields are created behind the check dam walls (figure 8-1). As the dam continues to collect alluvium, runoff is stored in the field in the form of soil moisture. An important principle operates in this technique: by capturing runoff from a broad catchment area and concentrating it in a reduced area, silt traps transform meagre quantities of rainfall (which otherwise would be lost to the production system) into utilizable soil moisture.

As shown in figure 8-1, four main types of trinchera wall design can be identified:

(1) piled rubble;
(2) stone alignments;
(3) stone facing with rubble backing; and
(4) double wall with rubble core.

This range of types reflects an increasing degree of engineering sophistication as well as a greater capacity of the structure to retain runoff. Trinchera walls were built so that they were buttressed against the valley walls and bedrock, thereby providing maximum strength.

Larger, more complex silt traps are now in common use in Mexico. These are called *atajadizo* and, like the smaller trinchera, they can be constructed out

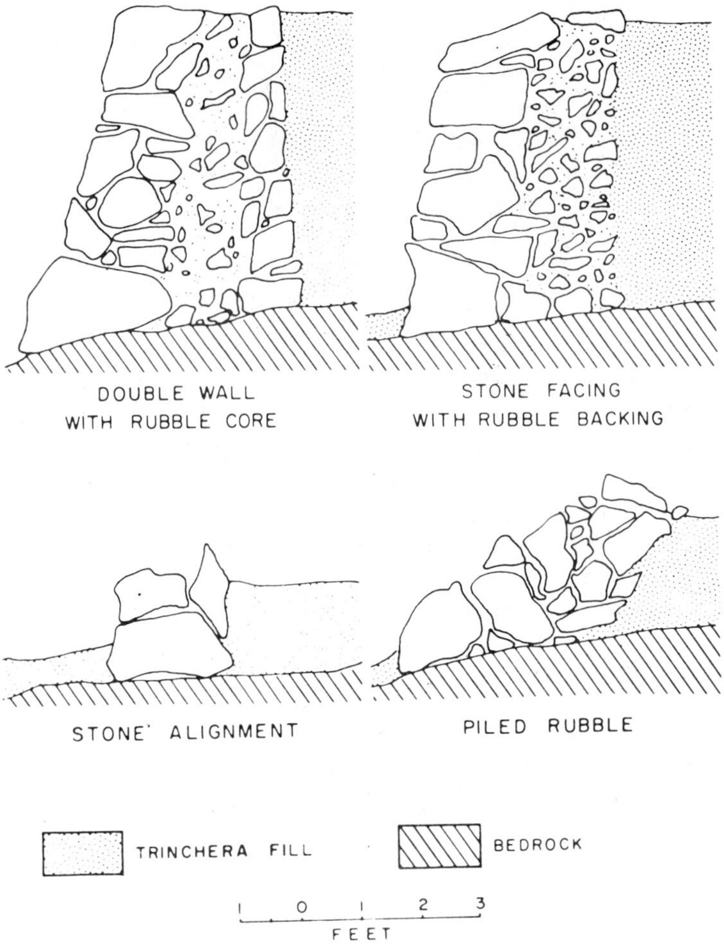

Figure 8-1 Representative cross sections of *trinchera* walls (Herold, 1970)

of rock. However, sometimes other materials such as earth, gravel or logs are used. The runoff and alluvium behind silt trap walls create level, flood-irrigated agricultural fields (figures 8-2 and 8-3).

On the basis of extensive interviews with Otomi farmers in Hidalgo, Mexico, Johnson (1977) was able to document the principles involved in successful silt trap *(atajadizo)* construction.

One of the most important considerations taken into account is the pressure that is exerted by the storm runoff on the structure. A poorly constructed atajadizo can be breached easily and washed away by runoff from a single storm.

Figure 8-2 *Atajadizo* silt trap in the Mezquital Valley, Mexico

Figure 8-3 *Atajadizo* silt trap series in the Mezquital Valley, Mexico

Well-constructed atajadizos incorporate one or more of the following features:

(1) double or triple stone walls separated by gravel rubble;
(2) walls that extend beneath the surface of the stream bed;
(3) an outer wall that is pitched upslope;
(4) a floodgate to release excess water; and
(5) curved walls allowing stormwater to be distributed evenly throughout the field.

Another important strategy used to protect atajadizos is to build them in a continuous stepped series along the length of the valley (*arroyo*). In this fashion, the total system of silt traps reduces the force of storm runoff and allows it to flow or seep slowly from one field to the next.

In sum, well-constructed silt traps impound water behind strong walls and allow it to seep slowly to the downslope field until the entire series has been watered. Individual silt traps (as well silt trap series) are built incrementally. As alluvium accumulates behind the walls, farmers build higher, more solid structures (figure 8-4). A farmer may tear down parts of the gully in order to enlarge the field. However, depending on gully configuration, the act of raising the wall alone will serve this purpose.

One important principle of silt trap construction in Hidalgo, Mexico, concerns the relative heights of the dam wall and the field behind it. Otomi farmers always advise that atajadizo should be kept 0.25 to 0.50 m higher than the field in order to impound storm runoff successfully and to allow it to infiltrate the soil. If alluvium is allowed to accumulate up the height of the wall, then storm runoff will overtop the structure, thereby depriving the field of necessary moisture.

These articulated principles reveal an important feature of silt trap agriculture: it is an intrinsically incremental and constructive system. The reasons for this are twofold. First, as we have seen, series of silt traps are always safer than single silt traps; and second, farmers must always add to the height of their structures in order to keep the latter above the level of accumulating alluvium.

The size of silt trap structures, as well as the size and shape of their associated fields, varies enormously. Johnson (1977) documented structures in one Mexican community ranging from 0.15 to 7 m in height; 1.5 to 19 m in length; and 0.1 to 2.5 m thick. The size of silt trap fields varies from less than two square metres to approximately four hectares. The shape of individual fields depends on gully configuration.

In order to assess the adequacy of silt traps one must consider two aspects: first, their cost; and second, their effectiveness in harvesting storm runoff.

Like contour terraces, silt traps require investment in human labour rather than capital. However, unlike contour terraces, silt trap walls require materials such as stone or logs which might not be located near the site. The

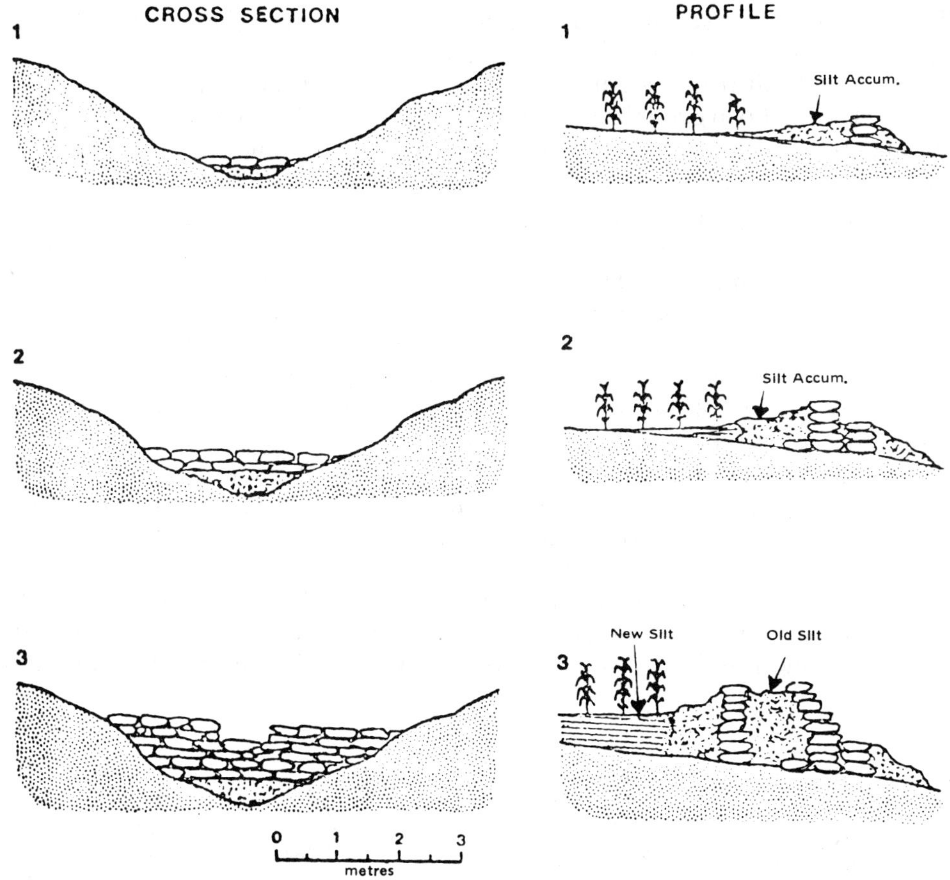

Figure 8-4 *Atajadizo* build-up (Johnson, 1977)

variable distances from which these materials must be hauled, as well as lack of field documentation make the labour costs of silt traps difficult to estimate.

One such estimate by Wilken (1976) calculated that one to two cubic metres of on-site rock could be excavated and placed in one working day. If the farmer must haul construction materials to the dam site, this will increase labour inputs considerably. Wilken concludes that distance is a crucial variable. If a farmer must travel as little as 200 metres for suitable rocks, this doubles construction time over what it would take if the rocks were available on site.

Even skilfully constructed silt traps are subject to weakening and failure and therefore must be monitored and maintained.

A well-maintained silt trap which is periodically built up as alluvium accumulates, can double in size in the space of a few years. On the other hand, silt trap systems are sensitive to inadequate labour input and can deteriorate rapidly resulting in severe erosion problems.

One such silt trap system in Nochixtlan Valley, Oaxaca was analysed by Spores (1969) who argued that following a drastic decline in population during colonial times, the extensive system was partially abandoned, resulting in dramatic, although not necessary irreversible, environmental deterioration.

In spite of high labour inputs relative to contour terracing, silt traps are a generally inexpensive and effective technique to create wet alluvial fields in sub-humid environments. The actual quantities of harvested water depend greatly on the meteorological and physical conditions of individual sites. Nevertheless, there is general agreement among those who have studied silt traps in contemporary Mexico that this technology results in productive and dependable (short of total rain failure) agriculture in regions which otherwise are unable to support farming.

The Mexican Government has incorporated simple silt traps into various soil and water conservation programmes. In one case, very rudimentary trincheras were used in a government-sponsored project in Chihuahua (figure 8-5). The aim of the project was to equalize stream flow throughout the year in one basin. Reports indicate that the 105 trincheras which were installed along the basin's stream courses resulted in a shift from an ephemeral to a quasi-perennial flow. Trincheras were found to slow down the exit time of water and

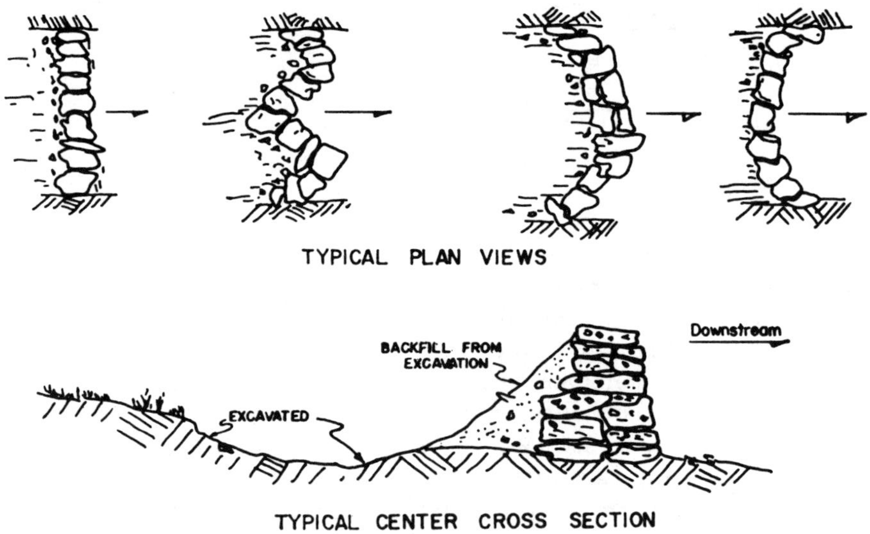

Figure 8-5 Plan views and cross section of conservation *trincheras* in Chihuahua, Mexico (Griffin and Dennis, 1969)

thus raised ground water levels and soil moisture. Among other benefits, this has actually enabled earlier plantings in this region (Griffin and Dennis, 1969).

CHECK DAMS

Check dams are small dams which impound storm runoff. They are a common feature of rural landscapes in many parts of the world. Check dams are built across gullies in China, India, Central America and elsewhere. The water harvested by check dams is generally used for livestock and irrigation, but if properly treated it can sometimes be used for human consumption.

Check dams may be relatively simple structures causing very little alteration of local topography. They may also involve considerable alteration or preparation of the catchment prior to water harvesting. Check dams of the first type are characteristically located at favourable sites, such as across water courses in narrow valleys with impervious rock or soil strata (figures 8-6, 8-7 and 8-8). These simple structures have the advantage of comparatively low cost, but the number of favourable sites available is usually limited. Where feasible, such check dams should be given priority.

The second type of check dam is more sophisticated and involves a greater input of labour and capital. The catchment and impoundment areas are usually considerably altered in this type of check dam through land alteration. This would include compaction of earth on the catchment, removal of stones and vegetation on the catchment to increase runoff and planting grass cover to increase runoff but decrease soil erosion. In some areas the catchments are chemically or physically treated to increase runoff and decrease seepage.

Check dams are an effective means of harvesting and storing storm runoff from large catchments, even under arid conditions.

Many contemporary check dams probably have been in continuous use since ancient times. In most fundamental ways, the structure and function of modern check dams are the same as their ancient counterparts. Thousands of small check dam reservoirs serve to harvest seasonal runoff to meet (partially or entirely) the domestic and agricultural needs of peasant communities throughout subhumid Mexico. Much like their ancient counterparts, these structures take advantage of local topographic features enhancing their water-retaining capacities with minimum investment in capital and equipment. Frequently, the labour required to construct and maintain these reservoirs is organized under traditional labour-sharing arrangements, thus building upon the ancient collectivist heritage of Mexico's rural communities.

Owing to the immense variety of site and situation and the absence of country-wide surveys, the overall adequacy of check dams is difficult to evaluate. However, the most commonly mentioned limitations include

Figure 8-6 A valley dam in Mt. Lofty Range, South Australia, showing the adjacent pump-house on the left, which supplies water to nearby dairy buildings. The bridged over-flow channel is just visible on the right

Figure 8-7 Check dam, Hidalgo, Mexico

Figure 8-8 Check dam, Hidalgo, Mexico

problems of unreliable and insufficient supplies, and poor quality. Frequently, communities must resort to a range of sources including wells, streams, springs, and reservoirs to meet the total needs. Among these, check dams are acknowledged to be the most vulnerable to contamination, easily becoming foci of water-borne diseases.

Of particular interest are the large check dams constructed during the eighteenth and nineteenth centuries. These massive structures (figure 8-7) were built to provide the water supply for haciendas, and thus are different in many ways from the smaller scale community check dams. Hacienda reservoirs owe much to design principles of European origin. Their masonry walls are extremely broad at the base (6 to 7 metres is not uncommon) and are provided with additional supporting buttresses. Lateral spillways are usually provided, and siltation problems are taken care of by sluice gates at the base of the structure.

The reservoir in figure 8-7 is one of seven similar structures providing one former hacienda in Hidalgo, Mexico, with irrigation water. Supplies are usually enough to irrigate 40 hectares (30 per cent) of the local cultivable lands. Additionally, reservoir silt provides an important source of renewable fertile topsoil for the farmers' maize fields.

Another example of a hacienda reservoir that is still in operation is located in another community of Hidalgo, Mexico. As in the previous example, the massive masonry structure (4 metres wide at the top and 35 metres tall, as measured on its downslope side) combines European design features with a strategic location at a point where the valley narrows significantly (figure 8-8). .

The purpose of this check dam is different from the previous example. Between August and February the dam's numerous sluice-gates are closed, allowing storm runoff to accumulate behind the structure. In February the water is released and farmers plant maize in the rich water-saturated alluvium of the valley bottom behind the dam. The crop matures early, benefiting from stored moisture and early summer rains. The gates are then closed in August to allow runoff to accumulate again for the following year.

As this valley drains an extensive watershed, farmers seldom experience a complete failure of storm runoff, and are able to cultivate a rich band of wet valley land extending several kilometres above the dam.

The only risks in this harvesting system take the form of early frosts and failures of the early summer rains. In order to harvest a crop before the reservoir starts flooding again in August, farmers must plant their maize early in spring during the time when occasional frosts can damage or kill the young plants. Sometimes when this happens farmers are able to replant in time. At other times, the frost comes too late for replanting. Accumulated soil moisture enables the young plants to survive the hot months of April and May; however, after this they require one or two rainfalls to reach maturity. Farmers here also attempt to enhance and conserve soil moisture by fashioning individual microcatchments for each plant.

CHAPTER 9

Rain and Stormwater Harvesting using Floodwater Farming and Inundation Canals

FLOODWATER FARMING is a technique used to harvest storm runoff by planting crops in areas likely to be flooded, either by channelled or sheet runoff. This is a risky form of water harvesting since crops fail in dry years (according to one estimate, at least two to three floodings are needed for a successful crop). Moreover, heavy rains, if not controlled by the farmer, can result in the destruction of the entire field. On the positive side, costs in terms of both materials and labour are extremely low.

This type of farming has made agriculture and settled life possible in many dry areas of the world. The technique has a history of considerable antiquity. However, owing to erosional and depositional changes in areas where floodwater farming was practised, hardly any artifacts of this technique remain. Until quite recently, floodwater farming resisted cultural and technological pressures for change and remained the most important water harvesting strategy of people living in dry areas in Africa, the Middle East, southwestern USA, Central America, India and Pakistan (figure 9-1).

Floodwater farming is still used on 73 per cent of the cultivated land of the Hopi Indians in southwestern USA. However, all evidence points to the fact that contemporary floodwater farming is now practised over only a fraction of the area that it once covered. For example, one recent study of the Hopi (Bradfield, 1971) estimates that approximately two-thirds of previously cultivated floodwater farmland have gone out of cultivation. Since the Hopi have been one of the main agricultural groups in the southwest this indicator becomes highly significant.

Nevertheless, in spite of its diminishing importance in southwestern USA numerous recorded examples of recent or current floodwater farming exist in the literature. The Zuni, a village-dwelling people in west-central New Mexico, have a long history of floodwater farming. Several tribes of the Gila River and lower Colorado River—the Mohave, Yuma, Cocopa and Maricopa— maintained floodwater irigation systems in which dams, ditches, dykes and walls were used to divert storm-swollen riverflow to areas in which crops were grown after floods receded, a system which must be similar to the inundation canals of Pakistan and India discussed below. The Navajo and Hopi of northeastern Arizona, the Papago of southern Arizona, and Spanish-American settlers in New Mexico utilize arroyo and slope floodwater for small-scale agriculture.

The most important decision floodwater farmers must make is selecting a site for their fields. Three principal types of site are preferred for floodwater farming:

(1) gentle slopes below escarpments;
(2) alluvial fans below valley mouths; and
(3) areas adjacent to streams and valleys, where waters overflow during heavy rains.

Each type of site involves somewhat different techniques, harvesting principles, and risks, and therefore will be treated separately in the following section.

Slopes Below Escarpments

Fields located on slopes below escarpments receive sheet runoff from the higher elevations. At these locations the catchment area is limited in size, but runoff is often high in proportion to rain. Thus, even small storms produce

Figure 9-1 Water-spreading dikes constructed in series in Pakistan to slow floodwater and harvest the water by seepage (Medina, 1976)

useful quantities of runoff. This is not necessarily the case with other floodwater harvesting techniques.

Fields receiving escarpment floodwaters do not usually include protective structures or water spreading devices. Therefore, farmers must select the sites carefully in order that flooding takes place without destroying the planting or burying it under loads of detritus.

Alluvial Fans Below Valley Mouths

Such fields are located where a watercourse draining runoff from higher elevations fans out upon reaching the more level ground of the valley floor. These fields are a favoured type of floodwater farming since runoff from the valley spreads out naturally over the fan surface without any need for artificial spreading and diversion. However, these fields do shift as the alluvial fan left by one flood may be redistributed downslope. Figure 9-2 shows this phenomenon clearly. In this example both natural and human factors control

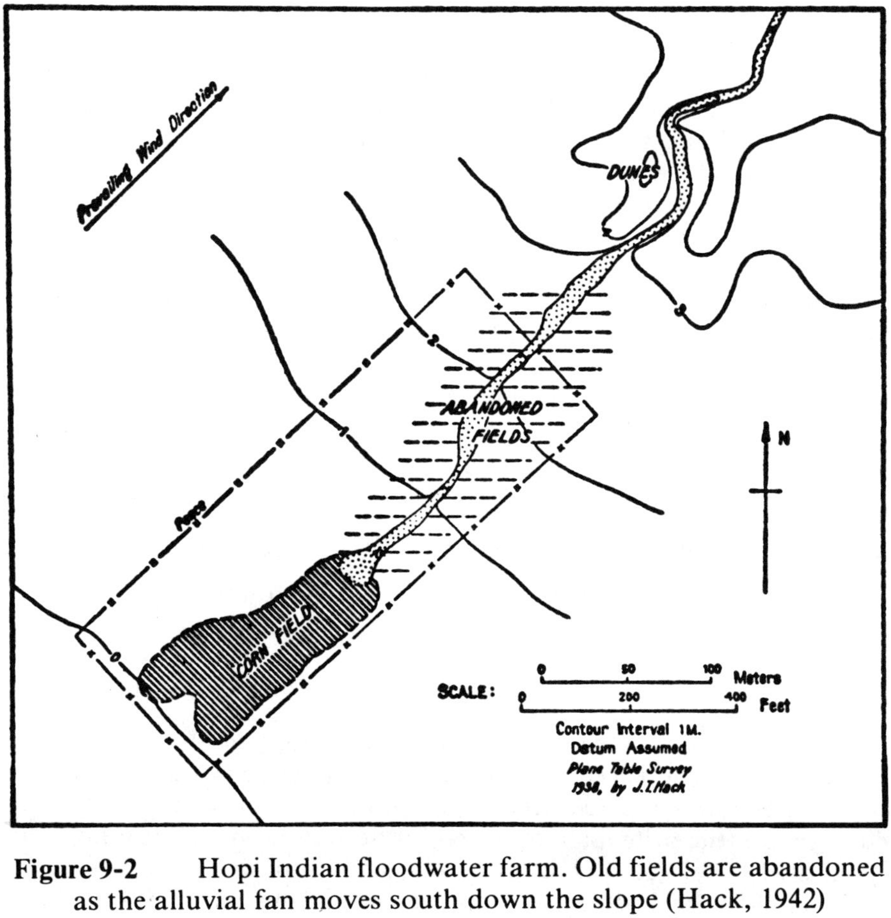

Figure 9-2 Hopi Indian floodwater farm. Old fields are abandoned as the alluvial fan moves south down the slope (Hack, 1942)

erosion. The natural factors include sand blown upslope by prevailing winds, forming dunes at the channel mouth. This assists in stabilizing water flow through the farm system.

Farmers also attempt to stabilize and control by constructing spreaders, dykes and channels (figure 9-1). A late nineteenth-century description of Zuni Indian floodwater farming in the USA indicates that low earthen spreaders were built at intervals for the entire length of the fan in order to ensure an equal distribution of water and alluvium to all parts of the fan. Other farmers were known to have reinforced earthen dykes with brush and stakes, and also to have carried soil to sections that needed to be levelled in order to ensure optimal water spreading. Today the Hopi Indians further aid water-spreading by digging channels to drier areas on the fan or, during droughts, by diverting water to individual plants (Bradfield, 1971; Castetter and Bell, 1942).

Light rains are of little use to floodwater harvesting since they do not produce sufficient runoff to bring about valley flow. One study in the USA covering a three-year period suggests that storm events with the required 50 to 75 mm needed to water these fields occur on the average 10 times during a summer growing season.

The uncertainty associated with unpredictable rainfall is lessened by locating fields on sites where heavy storms are rare, but where the fields will receive runoff from large catchment areas draining regions with a higher rainfall (or snowfall) frequency. Figure 9-3 indicates that most fields in southwestern USA depend on runoff from large catchments with the farmed area representing three to six per cent of the harvest watershed.

An additional important factor to consider here is the differential permeability of different catchments. Thus the volume of runoff can be more dependent on the type of surface than on the extent of the harvest watershed. One study (Bradfield, 1971) suggests that the principal land forms can be classified in three groups according to their runoff yield:

(1) highly absorbent, little runoff:	upper hill top, sand dunes, sand slopes.
(2) moderately absorbent, some runoff:	side valley slope.
(3) little absorbent, high runoff:	lower hill top and rock ledges, talus slopes, 'wash-down' slopes.

According to a recent estimate (Bradfield, 1971), one hectare of floodwater farmland is required to support one person, thus approximately five to six hectares are required to support one family. The crops grown include maize (the principal crop used for both household consumption and trade), as well as vegetables, beans, melons and squash.

Bradfield's 1971 study of floodwater farming in southwestern USA estimated that 360 hectares of akchin fields (from over 1,000 hectares at the turn of this century) were being cultivated at that time and that only 30 persons were engaged in full-time farming. The author calculated that these lands are going out of cultivation at the rate of 80 hectares a decade.

Among the most important forces contributing to the diminishing importance of floodwater agriculture are drier conditions which have led to the abandonment of the more marginal fields, and social forces such as the migration of young males and an increasing reliance on wage work and sheep-herding.

Floodplain Water Harvesting

Hopi and other southwestern USA Indian groups harvest storm-swollen stream water by diverting and spreading floodwater to fields located on low river terraces adjacent to valleys. This method is somewhat risky compared to other more dependable forms of moisture management.

Sites for this form of water harvesting are somewhat limited to portions of valleys and washes which are shallow enough to flood their banks, but which are not subject to violent floods.

The decline of traditional water harvesting in the American southwest is illustrated by the case of the Papago Indians of southern Arizona. The Papago, a semi-nomadic people inhabit a region receiving 125 mm of rainfall a year. Before colonial times, the region supported approximately 10,000 inhabitants

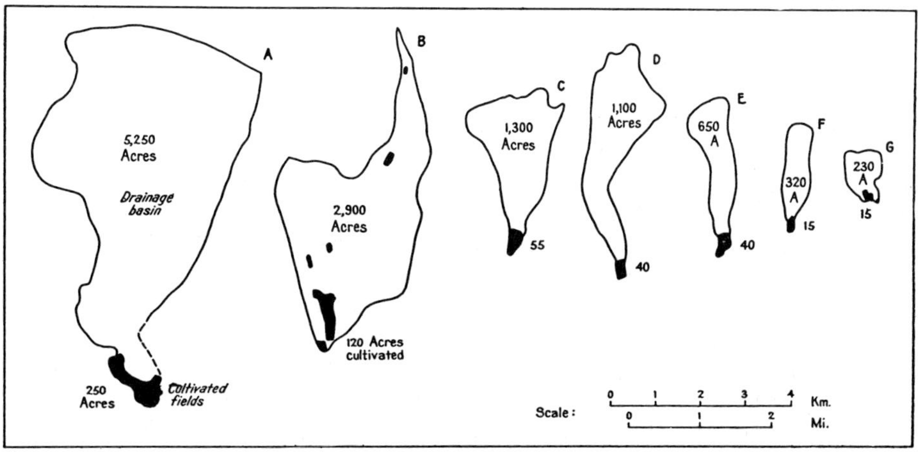

Figure 9-3 Cultivated floodwater fields in Arizona in relation to their watersheds (Hack, 1942)

who practised floodwater harvesting, hunting and gathering. An estimated 0.2 to 0.8 hectares provided families with 10 to 100 litres of maize a year, thus providing up to 20 per cent of their necessary food supply. In 1914, 5,662 Papago lived on about 360 hectares of floodwater cultivated lands, supplemented by stock raising, hunting, gathering, and wage work. At present, the Papago reservation is surrounded by 136,000 hectares of irrigated farmland. The Papago have mostly abandoned subsistence agriculture and seek work in irrigated fields. Additionally, the US government has built more reliable sources of water supply including deep wells, reservoirs and piped-in water, thus decreasing the need to maintain a water harvest system.

INUNDATION CANALS

Harvesting of rain and stormwater is carried out in semi-arid regions using canals. Either the canal harvests excess runoff (stormwater) only, or it acts as a direct receiver of rain runoff from a catchment. Such canals have been used for many years in the arid sections of northwest India and Pakistan to gather excess stormwater. Practically the entire state of Sind in Pakistan was dependent on such canals at one time. These canals are referred to as 'inundation canals', or 'pynes' and are of a very simple design. They are excavated at right angles to the bank of a river. No headworks are built across the river such as barrages or dams, and the bed of the inundation canal is left at a much higher level than the river bed. The normal low-level of water in the rivers in this region occurs in the winter. During periods of flood, water rises in the river until it reaches above the canal bed and the canal then starts to convey the water. When the flood subsides, the canal dries up. Irrigation in many parts of Sind is even now done with inundation canals drawing water from the Indus and its tributaries. They flow from April or May to September each year. The canals are from 3 m to 100 m wide, and from 1 m to 3 m deep.

The slope of the canals is less than that of the river, so that cultivable land between the canal and the river can be supplied by gravity. As a result, the velocity of water in the canal is less than that in the river, and silt in the flood water gets deposited in the initial length of the canal. It is not uncommon to see that the bed level of a canal has risen at the end of the rainy season by about 1.5 m (Sharma, 1949). A canal regulator (a gate) is often necessary to control flow in the canal and is usually provided about 5-6 km from the river. Silt removal is an annual necessity.

An interesting application of inundation canals was in vogue till about two centuries ago in Bengal and in the Thanjavur delta in Tamil Nadu (Willcocks, 1930). The floodwater with its silt was carried by the inundation canals to the fields in these regions, which had simultaneously a good rainfall, unlike the arid northwestern part of the country. The silt fertilized the fields; in addition fish which swam via these canals into local lakes and tanks fed on the larvae of mosquitoes and helped reduce the incidence of malaria. Because of the value of the silt and fish, the water itself was actually considered superfluous, and the canal generally flowed almost full. During the Afghan-Maratha War in the eighteenth century and the subsequent British conquest of India, these irrigation systems were destroyed, never to be revived.

In the Himalayas, from Kashmir at one end to West Bengal at the other, canals aligned roughly along contours are dug to draw water from hill streams or springs. These canals are known as *kuhls*. The length of the kuhl varies from 1 km to 15 km. They have a trapezoidal cross section 0.1 to 0.2 m^2 in area and carry a discharge of 15 to 100 litre/sec. Many kuhls collect rainwater and snowmelt running off the slopes above them, and occasionally one finds that the discharge increases along the length of a kuhl as a result. The discharge also varies with the season. A single kuhl irrigates an area of 80 to 400 ha through distributaries or by flooding. The irrigated land, being situated on hill slopes, is terraced. The system is common at altitudes from 350 to 3,000 m in the Outer and Middle Himalayas. Where there is a significant drop in elevation in the path of a kuhl, the fall is utilized to drive simple machinery like flour mills. The cost of construction of a kuhl varies from Rs3,000 to Rs5,000 per km.

CHAPTER 10

Rain and Stormwater Harvesting using Haffirs and Bunds

BUNDS

THE PRACTICE of putting small earthen dams (called *bunds*) across local streams to collect the rainwater running off the neighbouring region is a very old technique used in south India and Sri Lanka. The reservoirs so formed are called 'tanks' and the water collected is used for irrigation as well as domestic purposes (figure 10-1). The water-spread of these reservoirs varies from a few hectares to a few thousand hectares, depending on the nature of the terrain, rainfall, nature of the land downstream and runoff from the catchment. No effort is made to treat the catchment to increase the runoff, since the tanks are built in areas of moderate to heavy rainfall. There are about 500,000 tanks in India (Anon., 1966), a majority of which are found in peninsular India. A large number (about 100,000) of small tanks are found in west Bengal, but they differ from the tanks of peninsular India in that they are excavated in the ground, rather than formed by obstructing a stream.

The ideal site for a tank satisfies the same conditions as for any other reservoir including a reasonably deep gorge at the site so that the length of the bund is short. Also required is a local source of good soil to build the bund. Since the quantity of water stored is small, one cannot afford to lose much of it in conveyance, and hence the bund should be located near the place of use, e.g., a village or an area to be irrigated. Sometimes, when a stream or valley is large, many tanks are built in series so that water flows to a lower tank only after the upper ones are filled. In this way, the bunds act like check dams (chapter 8) or the gully dams of China (chapter 11).

Bunds which are built of soil will have a top at a fixed level. If floodwater flows over the bund, the bund will breach rapidly due to erosion. Therefore, a waste weir is provided, which is simply a masonry wall of suitable length whose top is at a level lower than the top of the bund. Excess water will escape over this waste weir, and hence will be conducted directly to the lower reaches of the stream through a channel. The top level of the waste weir therefore is the 'full tank level' and determines the amount of storage in the tank. The waste weir may be built alongside the bund or at any other suitable location around the perimeter of the tank. The maximum flood likely to occur in the stream during

Figure 10-1 Tanks near Bangalore in India

the life of the tank should pass over the waste weir, so that the maximum water level (during flood) is between the top of the waste weir and the top of the bund. The difference between the maximum water level and the top level of the bund is the 'freeboard'.

In northeastern India, a tank usually caters for both domestic and agricultural needs; occasionally fish are cultivated in tanks. Water for domestic purposes is manually transported in pots, and therefore steps are provided near the deepest part of the bund, since water is available there for longer periods than in any other part of the tank. Water for agricultural purposes must be transported through a system of channels with discharge controlled by a gate or a plug valve.

The older bunds in India can be massive and were made entirely out of locally available mud. They are trapezoidal in section with flat (1:3) side slopes. Huge boulders were often used to protect the upstream slope from erosion (figure 10-2). Modern tank bunds, however, have a cross section similar to that of earth dams. They consist of a 'hearting' made of impervious earth, such as clay; a 'casing' made of local earth; a 'cutoff wall' again of impervious earth, taken deep enough into the ground to meet the impervious stratum underneath; and 'drains' to lead seepage safely across the bund. The hearting, cut-off wall and the drains all serve to keep the downstream part of the bund from becoming wet, as otherwise the seeping water will erode the bund from the rear. Figure 10-3 also shows the 'seepage line' which is the boundary of the wet part of the bund. The rear slope of the bund is covered with turf to prevent erosion by rain. The front slope (on the water side) is protected from sloughing and erosion due to waves by means of stone pitching from the bottom up to the expected level of the waves above the maximum water-level.

The bund is built using human and animal labour and, as a rule, no

Figure 10-2 Upstream slope of a large older tank. Note protective stonework

machines are used. The earth for the casing is excavated at some distance upstream of the bund, transported by bullock carts, and dumped in layers throughout its length. The casing and hearting are simultaneously built up. The earth is automatically compacted as construction progresses, due to the walking of numerous people and animals. As a bund takes two to three years to build, the rains during the period also help in compaction.

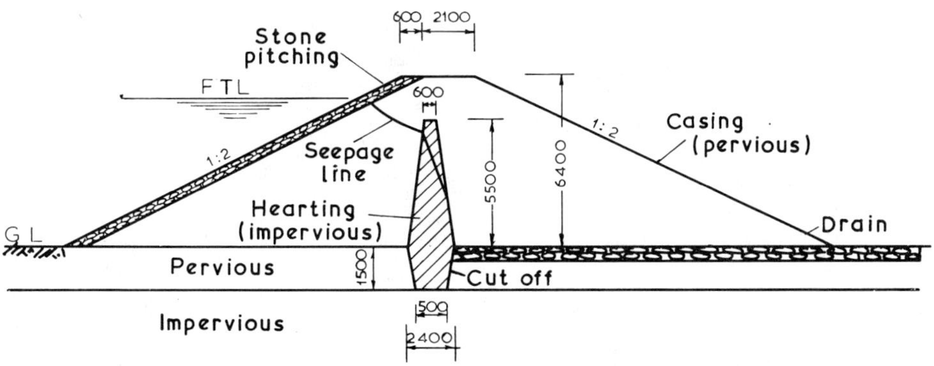

Figure 10-3 Cross section of a bund

Irrigation water is taken from the tank through an outlet tunnel across the bund (figure 10-4). The flow is controlled by a gate (or a plug valve in the case of small tanks) operated by a lead screw installed in the upstream side of the bund. Approach to the operating platform is over a small bridge. The tunnel is lined with masonry, and it is important to ensure a good watertight joint between the masonry and the earth to prevent erosion of the bund at this joint.

The waste weir is a very important component of the tank and should be designed to discharge the maximum flood likely to occur during the life of the tank without the water level rising above the designed maximum. In many tanks, the waste weir is situated at an extremity of the bund, so that the amount of masonry construction is small. Sometimes it is level with the ground and in rocky areas, excavation of the rock is done by blasting to form the waste weir (figure 10-5). Where the top of the waste weir is above the ground, a rather flat slope paved with stones should be provided on its downstream side (figure 10-6), so that the surplus water flowing out does not scour the ground below the bund.

Tanks used as an irrigation supply must hold a reasonably large quantity of water. On the other hand, the bund should not be too long, so that it will be economical to build. From the same consideration, not too much land should be submerged by the water which fills the tank. The terrain best suited for tanks is of undulating nature, with hillocks or raised mounds in the gaps

Figure 10-4 Tank outlet consists of a deep vertical tunnel made of stone
(Inset) Cross section of plug valve and outlet tunnel

between which short bunds could be built, with fairly steep slopes so that a large quantity of water could be stored below a given contour. Such undulating terrain contains numerous watersheds, bounding small catchments, so that the flood to be discharged by waste weirs of tanks in a given catchment will not be too heavy.

Figure 10-5 Waste weir excavated by blasting an area in rocky substrate

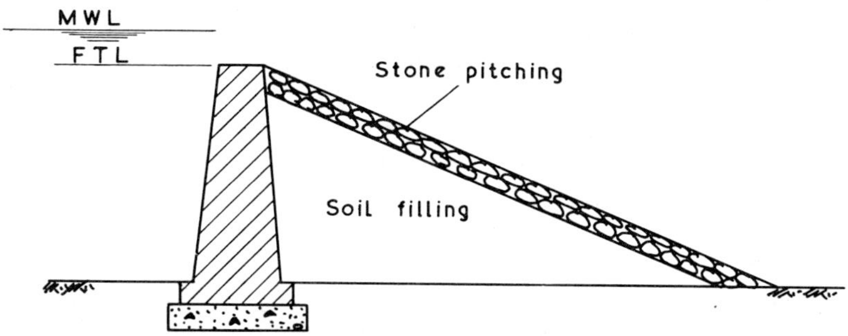

Figure 10-6 Section of a waste weir. Flow of excess water will occur over the paved facing stones

Table 10-1 gives the area irrigated by tanks in relation to the total irrigated area, and also the density of tanks in various states of India in 1963. Although the absolute magnitudes may change over the years, the relative values will still be valid. It can be seen that around 40 per cent of the total irrigated area in the eastern states in India is served by tanks. In many other states, the contribution from tanks is also significant, though not to the same extent as in the eastern states. Individual tanks have storage capacities which vary widely from one region to another, e.g., in Karnataka some store as little as 200,000 m^3, while in Maharashtra the minimum capacity of tanks is 1.4 million m^3. There are even tanks in Tamil Nadu which store as much as 100 million m^3.

The exact total volume of water collected in tanks is not available, but a rough estimate made from the area irrigated of 4.55 million hectares gives an average storage of 351,700 m^3 of water per tank. The average area irrigated per tank works out to 9 ha. The total amount of rainwater stored in the tanks of the three southern states of Andhra Pradesh, Tamil Nadu and Karnataka amounts to 9.7 per cent of the annual rainfall over the area.

Silting is a serious problem with tanks. Tropical rains often consist of sharp showers, and the hot weather preceding the monsoon tends to loosen the soil particles. Thus, considerable erosion takes place, especially in cultivated areas. Tanks are almost always in the midst of such cultivated areas, since the irrigated area must be situated near the tank in order to minimize loss of water by seepage from delivery canals. The rate of erosion of soil in India at large reservoirs (Anon., 1978) is estimated to be of the order of 600 to 900 m^3/km^2 every year. This is carried as silt deposited in tanks or carried into the sea. For large catchments, the erosion could be as high as 2,250 m^3/km^2/year (Murthy, 1978). On an average, the storage capacity of a tank decreases at the rate of about 0.5 per cent per year due to silting (figure 10-7). The outlet to the irrigation canals must therefore be kept at a level which allows a 10 per cent 'dead storage' which will ultimately be filled with silt. To prevent silting, afforestation of the catchment of the tank, banning cultivation of the foreshore lands and construction of small silt traps upstream are recommended. Silt traps could also later provide cultivable areas of their own (chapter 8).

Breaching is the most serious problem with tanks. A case study in Andhra Pradesh (Anon., 1966) revealed that 20 tanks included in the study had suffered a total of 38 breaches. Of these, 14 were due to the bund being overtopped by floodwaters, 16 were due to erosion of the bund, 5 were due to 'piping' where seepage water flowing below or through the bund took away soil particles when emerging to the surface, and 3 cases were due to unknown reasons.

Most cases of overtopping of bunds occur due to insufficient capacity of the waste weir. In the above case study, the committee concluded that on average, the capacity of the waste weir was only 23 per cent of the estimated maximum flood. Sometimes local farmers, in order to increase the storage in the tank, raise the level of the waste weir crest before the monsoon, but do not

Table 10-1 Water harvested by tanks in India 1963

State	Population, millions	In thousand hectares		Contribution of tanks to irrigation, %	Total land area, thousand km^2	Total number of tanks	Tank density per km^2
		Area irrigated from all sources	Area irrigated from tanks				
Andhra Pradesh	43.5	2,967	1,232	41.5	276.8	56,700	0.2
Assam	14.6	613	0	0	78.5	0	0
Bihar	56.4	1,973	276	14.0	173.9	27,800	0.16
Maharashtra	50.4	1,080	191	17.7	307.8	48,100	0.16
Gujarat	26.7	720	16	2.2	196.0	20,400	0.10
Jammu & Kashmir	4.6	280	0	0	222.2	0	0
Kerala	21.3	336	45	13.4	38.9	1,500	0.04
Madhya Pradesh	41.7	952	151	15.9	442.8	40,200	0.09
Karntaka	29.3	883	355	40.2	191.8	36,500	0.19
Orissa	21.9	1,028	412	40.1	155.8	1,700	0.01
Tamil Nadu	41.2	2,460	928	37.7	130.1	31,400	0.24
Punjab	13.6	3,154	6	0.2	50.4	200	0.004
Rajasthan	25.8	1,790	165	9.2	•342.2	Not available	
Uttar Pradesh	88.3	5,013	413	8.2	294.4	140,000	0.48
West Bengal	44.3	1,336	364	27.2	87.9	100,000	1.14
Total			4,554			504,500	

Source: Anon., 1966

restore it before the floods. This could result in overtopping of the bund. In the case of a 'series' of tanks, where many tanks are built across the same stream, the overflow from the upstream tank filling the one on downstream, the breaching of an upstream tank is likely to cause successive breaching of the downstream tanks. This risk can be reduced by making the downstream tanks bigger than the upstream ones and, of course, providing adequate capacity for all tanks. This topic is again considered in chapter 11 in relation to the Chinese techniques of flood control.

Tanks (unlike reservoirs) are usually shallow and thus their stored water covers a larger area relative to volume. This large area in relation to the storage in tanks means a relatively large evaporation loss will occur from tanks. Evaporation rate depends on atmospheric humidity, temperature and wind speed. In India, the hottest period of the year is just before the monsoon, when most tanks are almost dry. The windiest months are the monsoon months, when the humidity is also at a maximum. As a rule, the mean evaporation loss is assumed to be 1.8 m per year. While the actual evaporation is close to this

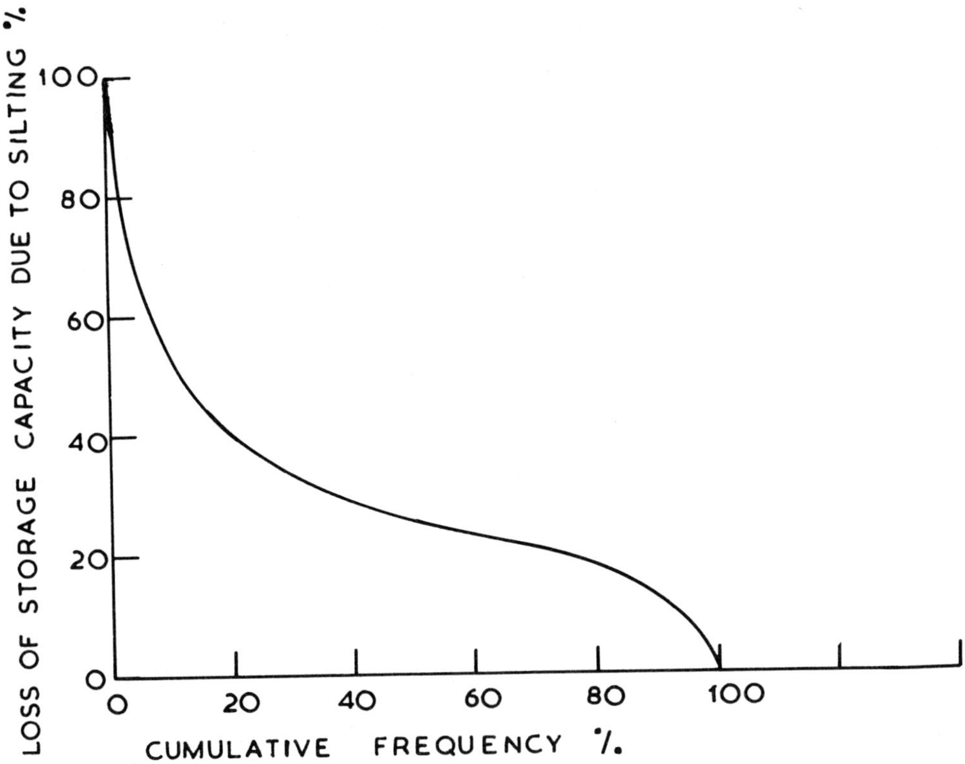

Figure 10-7 Cumulative frequency distribution of silting of tanks in Tamil Nadu

figure in Tamil Nadu or Andhra Pradesh, it may be as high as 3 m per year in the hot arid regions of northwestern India.

No satisfactory method of reducing evaporation has yet been found in India. Planting of trees on the shores of the lake has been suggested to reduce the wind velocity on the lake surface, but obviously more research is needed, similar to that reviewed in chapter 5.

Very little data are also available in India regarding the loss of stored water by seepage through tank beds. A recent study revealed that the average rate of seepage from the bed of a tank in Aurangabad district in Maharashtra state was 0.3 m/month. However, this rate is unlikely to be maintained as the tank grows old, since fine silt is deposited, the bottom mud becomes dense, and progressively the rate of seepage is cut down. Other sources estimate that 1.2 m of water percolates into the ground through tank beds every year (Anon., 1966).

Tanks need annual maintenance after every monsoon. Weeds growing along the top of the bund or waste weir are pulled out, eroded parts of the bund repaired, ratholes plugged, clogged drains cleared and the upstream revetment stones properly packed. When the tank dries up before the monsoon, accumulated silt is cleared. This, however, is easier said than done. In the earlier days, farmers who cultivated lands adjacent to a tank used to dig out and transport the silt and spread it on their lands as manure, because silt is estimated to have a nitrogen content of about 0.3 per cent. With rising transport costs and the advent of chemical fertilizers, this practice is often now abandoned, resulting in more rapid silt accumulation in tanks than before.

Bunds are likely to settle during the first few years after construction. The level of the top of the bund should therefore be maintained by periodic filling. Any cracks in the bund should be plugged with a mixture of sand and clay, especially at the joint between the earth and masonry. Cracked joints in the waste weir masonry should be repaired. Any scour by the lead-off channel at the bottom of the waste weir should be filled, so that the weir foundation is not undermined.

The current cost of construction of a complete tank system, including the bund, waste weir, irrigation canals, etc., has been estimated in terms of the area irrigated by it at $1,000/ha (Subbiah, 1978). Assuming an average total storage requirement of 39,100 m^3/ha, the cost of a tank works out to $0.25/10 m^3. The water is sold to farmers not on the basis of the actual volume consumed, but on the basis of the crop grown and the area irrigated. The average annual return works out to about $0.01/10 m^3 of stored water, which is about 4 per cent of the initial investment. Allowing for depreciation and maintenance costs, it is obvious that there is a large element of subsidy involved in the sale of water. But this is not peculiar to tanks, and is true of all irrigation works in India, though the subsidy is smaller with conventional large reservoirs due to economies of scale.

HAFFIRS

Haffirs are essentially excavations made at the bottom of natural catchments, thus they require modification of the site. In the case of bunds excavation is avoided except to provide the soil for the bund. Haffirs are commonly used in the Middle East and western India. In Africa they are found in many different countries, for example, the Ministry of Mineral Resources and Water Affairs in Botswana have indicated that this form of rainwater harvesting is practised by the rural people in eastern Botswana which has a characteristic summer rainfall. The Ethiopian Water Resources Authority note that this process of water harvesting is practised within the central highlands and some parts of the Rift Valley in Ethiopia.

Haffirs have been particularly successful in southern Sudan. For example in the vast southern plains area the soils are poorly drained with only a slight gradient. This results in an overland flooding during the rains with a 'creeping flow' of water across the region. Haffirs are therefore built by digging out a low area and the excavated soil is heaped up on the downslope to form a large bund. Water flows into the haffirs from the creeping flow and provides a dry season supply for the local cattle (figure 10-8). As a result of haffirs herdsmen tend to stay near their villages rather than to seasonally migrate and they are also in a better position to plant staple food crops prior to the early rains.

Normally, the haffir water is used mainly for livestock where boreholes are provided by the government for potable water. In areas where there are no boreholes, instead of local people taking water from the haffirs, it may be profitable to provide one section of the haffir with a sand-filled water storage tank and a well from which potable water can be drawn.

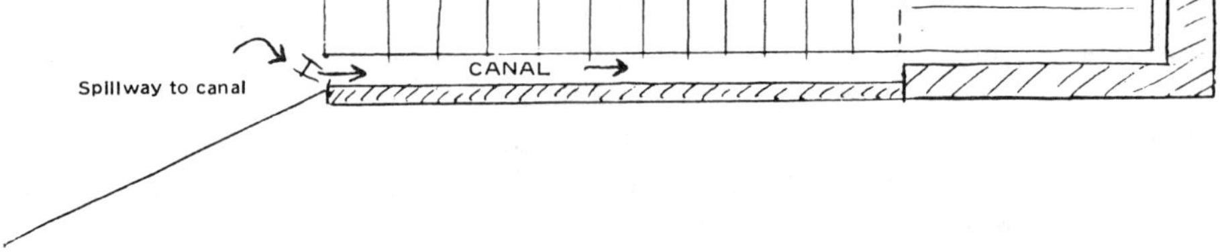

Figure 10-8 Ranching scheme using paddocks with water supplied as a *haffir* system

There are other variations in the southern Sudan, where it is possible that fast-growing rice varieties could be cultivated in paddies filled by harvesting the creeping flow (figure 10-9). Such systems when used in connection with water storage in haffirs seem to hold much promise for integrated development. They have the potential for providing water for nomadic tribes in areas where farming can be successfully demonstrated on a year-round basis.

The concept of a haffir is also used in a modified way in India. This is particularly seen where tanks are built by excavating the ground, and not by constructing a bund. These dug tanks are found primarily in West Bengal and Bihar State, and are used to rear fish. Other states where such dug tanks exist are Orissa, Assam, Manipur, Uttar Pradesh and Madhya Pradesh. The tanks collect rainwater running off the ground above them, and once filled, they are then sealed from rainwater on all four sides by embankments (figure 10-10). It is estimated that 4 to 10 hectares of catchment area is required to fill one hectare of tank bed.

Fish tanks are best built in non-rocky sites with at least 2 m deep topsoil. The water table should be high, so that too much water is not lost by seepage (the rate of fall in water level over a year should be less than 1 metre). The catchment area should be covered with vegetation, so that silt does not enter the tank along with rainwater. Another system similar to a haffir is the use of *ahars* in India.

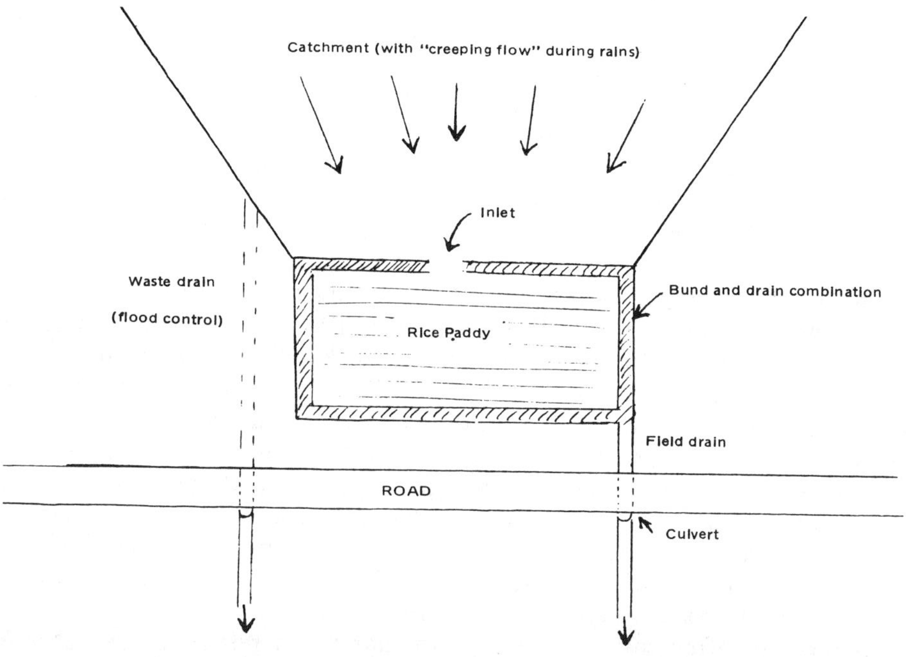

Figure 10-9 *Haffir* arranged to provide water for rice paddy in southern Sudan

Figure 10-10 Dug tank for fish cultivation in Bihar State

Ahars are created by building earthen bunds aligned as far as possible along contour lines (figure 10-11). Since the ground where they are built has a very mild gradient (the fall is of the order of 10 cm in 1 km), the bunds must be transverse to the contour lines. Since variations in ground elevation over short distances are very small, the ahar can follow local features like roads and property boundaries without any problem. During the monsoon, rainwater running off the ground above the ahar is collected in the ahar, and due to the flat nature of the ground, vast areas of land are submerged. After being retained till the sowing season all the water is then let out downstream through pipes embedded in the ahar at the gate (figure 10-12). The water can also be pumped to higher land for pre-harvest or pre-sowing irrigation (figure 10-13).

The soil which was earlier under water in the ahar is then sown (usually with wheat). The soil in these regions has a high clay content, so that its moisture retention capacity is high. The crop therefore grows up quickly without further watering. The 'submergence tanks' of Rajasthan and Uttar Pradesh also work on the same principle. Here the collected water is let out by the end of September so that sowing can be done by the middle of October. In Rajasthan's desert district of Bikaner, similar structures are built, but the water is used almost exclusively to soak the submerged land.

Ahars are often built 'in series', so that when released the water (or surplus stormwater) will drain out and fill another ahar downstream. Sowing on the bed of the downstream ahar is therefore delayed but the area of land

Figure 10-11 An *ahar* near Patna in India

Figure 10-12 Gate of an *ahar*

Figure 10-13 Irrigation outlet from an *ahar*

benefited by this practice is large. As in the tanks built in series, breaching of an upstream ahar in a series is likely to endanger the downstream ahars.

An interesting benefit resulting from ahars is that brackish groundwater in the neighbourhood of ahars often becomes potable after the ahar is built. This occurs because fresh water collected by the ahar easily percolates into the ground through the soil broken up every year by ploughing. Ahars also check soil erosion by arresting the flow of rainwater, and because of the numerous outlets from an ahar, there are generally no silting problems.

Sometimes in India a particular variety of rice is sown in the bed of the ahar before it actually fills up. This variety can grow in standing water, and although the yield is low, the effort is worthwhile, especially because a second crop of winter wheat will be taken once the ahar is drained.

Unlike the bunds of tanks, the ahar bund is much simpler. It does not contain a hearting or a cut-off trench, and the soil in the bund is the same consistency throughout its section. The section of the bund is trapezodial, with an upstream slope of 1:2 and a downstream slope adjusted to keep the phreatic line within the bund. This is not expensive, because the height of the bund generally does not exceed 3 m and the soil (because of the high clay content) has a low permeability. The top width of the bund is normally kept at 1 m. A freeboard of 1 m is provided to allow for waves. The length of the bund may vary from 150 m to 10 km or more depending on local requirements. The bund of the ahar near Patna (figure 10-11) follows the curves of the road to its right, and can be seen disappearing into the distance at the far left of the photograph. The bund slopes are usually left unprotected, but in deeper portions, boulder pitching can be done on the upstream slope. The top of the bund is sometimes used as a road. While building the bund, gaps are left to accommodate the masonry work involved, such as irrigation outlets and waste weir to dispose of surplus water.

The waste weir of an ahar is similar to that in the tanks discussed earlier. In addition, sluice gates set in masonry structures (figure 10-12) are provided at one or more points to empty the ahar quickly, if required, in time for sowing. The gates may also be operated along with the waste weir during flood control. The gates are made of steel plates stiffened with angle irons, and are operated by a screw. Their size is of the order of 1 m × 1 m. The surplus water is channelled to a local stream or another ahar downstream through an irrigation off-take.

The off-take from an ahar is very simple, consisting of a pipe embedded in the bund. The bund is protected from erosion at the ends of the pipe by small masonry blocks built around the pipe (figure 10-13). This pipe has a diameter of 150 to 300 mm and is laid at ground level. Since the bottom width of the bund is typically about 15 m, the length of the pipe is of the same order. The pipe may be made of stoneware, asbestos cement, reinforced concrete or cast iron. When water is to be retained in the ahar, these pipes are plugged at the end with straw and earth.

In India the land irrigated is generally adjacent to the ahar, thus the ahar canals themselves often act as field channels in which case there is no hierarchical network of main canals, distributaries and field channels as in the case of ordinary irrigation tanks. The ahar canals are of small capacity and closely spaced. The typical spacing between canals is 50 to 100 m and each canal has its individual intake. If the ahar feeds another ahar downstream, there will generally be no irrigation canals, but in a big ahar, apart from the small channels, several large canals may feed distributaries and field channels.

The main parameter indicating the utility of an ahar is the extent of the land submerged by the water stored in it, the actual volume of storage being of secondary importance. A flat terrain having a very mild slope is therefore ideal for the construction of ahars. As in haffirs the soil must be moisture-retentive, since the moisture absorbed when the ahar bed is under water should suffice for the crop till it is harvested. A sufficiently high clay content ensures this.

No information is available on the total quantity of water harvested by ahars. Individual ahars may submerge an area of 1 ha to 500 ha. Very large ahars (or 'submergence tanks') can submerge up to 4,000 ha, and in addition the stored water can irrigate an equal area or more, or be used to feed more ahars downstream depending on the mode of operation of the ahars. The quantity of water stored in an ahar may range from 5,000 m^3 to one million m^3, and in very big schemes up to 70 million m^3. The thousands of ahars existing in Bihar State irrigate a total area of about 800,000 ha (Anon., 1966). Ahars are actually more efficient in the use of stored water than irrigation tanks, since:

(a) the percolation losses are smaller due to the low permeability of the soil as well as the absence of an extensive network of channels;
(b) evaporation losses during storage are small since the humidity in the atmosphere during that period is high; and
(c) all the water required for the crop is stored in the soil, and little of it is lost by evaporation.

In India ahars are amazingly free from most of the problems encountered in ordinary tanks. The only serious one to be faced in ahars is breaching of the bund. This is due either to water overtopping due to insufficient disposal capacity or to breaching of an upstream ahar. The only answer to this is to provide a long enough waste weir and ensure that the sluice gates are opened in time. A less serious problem is erosion of the unprotected slopes of the bund. The slopes are left unprotected since water stands against the bund for only a few weeks. Erosion therefore occurs either by rain or by waves.

The bund in an ahar should be repaired every year after the monsoon, the eroded portions being filled and the levels restored where any settlement has taken place. The gate, if provided, should be painted periodically. Cracks in the masonry should be repaired.

Due to the absence of large networks of canals and bund slope protection measures, the cost of an ahar is much smaller than that of tanks for a given storage. Another factor which accounts for the low cost of an ahar is that only the land occupied by the bund itself needs to be acquired, whereas for tanks, the land submerged by the water at the full-tank level must also be acquired.

Accurate cost figures for constructing an ahar are not readily available since very few new ahars are being built. The thousands of existing old ahars were built by individual landowners over centuries. A large number of these are out of use, most of them due to breaching. These ahars are now being repaired by state governments. The cost of repairing the old ahars works out in the range of \$1 to \$3 per metre length of bund, based on the figures for 1963 (Anon., 1966). On the basis of the area benefited by irrigation, the total cost of repair will be \$5 to \$6 per ha. The cost of a new ahar is only slightly more than this since the repairs on which these figures are based are quite extensive. In terms of the volume of water stored, the cost averages from \$0.01 to \$0.02 per m^3 depending on the size of the ahar.

CHAPTER 11

Rain and Stormwater Harvesting in China

RAINWATER STORAGE

IT IS SIGNIFICANT that China with its vast population is actively promoting rain and stormwater harvesting in rural areas. Although examples from China have been cited in previous chapters, we must now take up the many examples in the proper context, that is, as they are used in China in an everyday setting.

Wells

Water-storage wells are below-ground structures for storing rainwater. These have been widely used in the Loess Plateau regions of China. For example, in Huguan County, Shanxi Province, more than 45,000 water-storage wells were constructed from 1970 to 1974. In many villages each farmer has his own water-storage well, and the wells play an important role in meeting the domestic water needs and are also used for small-scale irrigation.

Water-storage wells may be jar-shaped, bottle-shaped, or kiln-shaped (figures 11-1, 11-2 and 11-3).

A water-storage well consists of three parts:

(a) *Inlet:* including a water collector, silt basin and inflow pipe;
(b) *Dry shaft:* composed of the mouth of well and delivery section, which are the portions above the water storage itself; and
(c) *Well chamber:* the storage section.

In order to prevent seepage losses, the wall and the bottom of the water-storage well are provided with a waterproof layer. The bottom of the well is also protected against erosion by a bed layer.

The well should be located where it will have a certain catchment area for collecting sufficient rainwater. The well is usually situated near farmyards, roads, barren slopes or houses. It should not be situated too near buildings, otherwise it will endanger the building if the water-storage well collapses. The well should be constructed in hard soil, preferably in cohesive loess of low permeability. Table 11-1 shows the soil characteristics of water-storage wells in various counties.

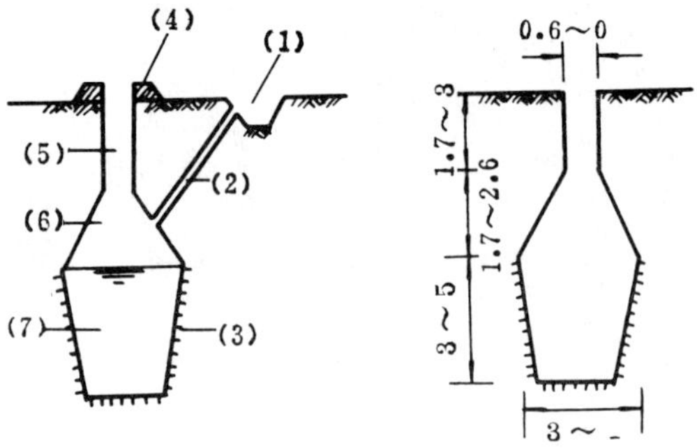

Figure 11-1 Cross section of a jar-shaped well in Gansu province. (1) silt basin; (2) inflow pipe; (3) holes for impervious layer; (4) platform; (5) mouth of well; (6) transition section; (7) well chamber

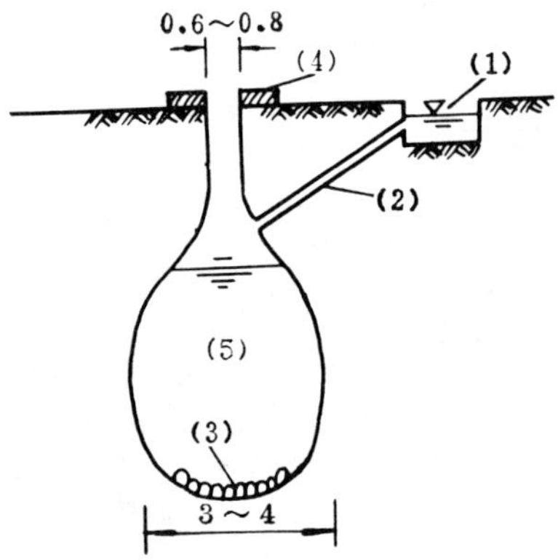

Figure 11-2 Underground bottle-shaped well for storing rainwater collected from courtyard areas in Shanxi province in China. (1) silt basin; (2) inflow water pipe; (3) protection bed layer; (4) platform at mouth of well; (5) stored water

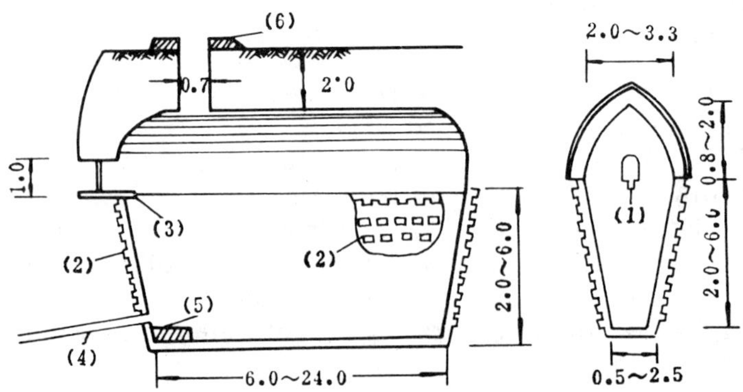

Figure 11-3 Cross section of a cave. (1) entrance; (2) holes for impervious layer; (3) intake flume; (4) release pipe; (5) protection basket; (6) platform

Table 11-1 The Soil Characteristics of Water-Storage Wells

Location: Province, County	Natural State				Soil gradation (percentage)			Uniformity Coefficient	Soil Type
	Dry Density (t/m^3)	Moisture Content (percentage)	Porosity (percentage)	Specific Gravity	Sand	silt	clay		
Shaanxi Suide	1.44	6.0	48.5	2.70	29	60	11	8.9	Light silty loam
"	1.68	13.0	37.8	2.70	19	54	27	7.3	Heavy silty loam
"	1.45	10.0	45.3	2.70	30	64	6		Heavy silty sandy loam
Shaanxi Yulin				2.71	30	54	16	12.3	Medium silty loam
Gansu, Yuzhong	1.58	20.0	41.7	2.71	12	66.5	21.5	12.0	Heavy silty loam
"	1.14	13.9	57.8	2.70	17	66	17		Medium silty loam
Gansu, Dingxi	1.27	7.7	57.0	2.74	21	60	19		Medium silty loam
Shanxi, Linshi	1.30	12.8	51.2	2.68	18	59	23	22.0	Heavy silty loam
"	1.33	11.6	50.3	2.69	26	59	15	22.4	Medium silty loam
"	1.27	13.5	52.8	2.70	37	52	11	10.7	Light silty loam
Henan, Jiyuan Gan	1.53	15.3	44.2	2.74	16	59	25		Heavy silty loam
Gansu, Dingxi	1.26	12.0	58.7	2.73	19	54	27		Heavy silty loam

The number of wells required in an area is based on the amount of precipitation and local catchment conditions.

The construction methods for the wells in Shanxi Province will be described by way of example. The cross-section of a typical local well is shown in figure 11-2. The mouth of the well usually has a diameter of about 0.6 to 0.8 m and a depth of about 1 m. It is finished with a smooth surface and fairly accurate dimensions. In between the mouth and the well chamber, the delivery section has a depth from 1 to 4.5 m. This secion should be excavated in a way that for every 0.3 m depth, it is enlarged laterally by about 0.1 to 0.3 m, thus resulting in a bell-shape. The well chamber, beneath this section, has a depth from 8 to 10 m. At the base this chamber is narrower, ending in a round bottom. After the well shaft has been excavated, it is covered at its mouth with a wooden plank to prevent cracking due to drying and weathering.

The silt basin is usually 1.0 to 1.5 m long, with a depth of 0.5 to 0.7 m. The bottom of the basin should be lower than the entrance of the inflow pipe. This pipe is constructed only when the well chamber has been excavated to a depth permitting the free-standing of a man in it. Then, after completion of the inflow pipe, digging may continue down in the well chamber.

This inflow pipe may be made of tile, earthenware or bamboo. Its gradient should be steep enough so that the water will fall directly on to the well bottom and not erode the wall of the well. The pipe should extend out of the well wall by a length of about 0.1 to 0.2 m, and it should be grouted so that the pipe is fixed tightly in position. At the opposite end, at the entrance, the inflow pipe should be equipped with a wooden trash rack or with a wire mesh.

The construction quality of the impervious layer of the well exerts a direct influence on the storage of water; therefore it should be done carefully.

In regions where lime is available, a lime-clay mixture is the best choice for this impervious layer. The well wall is compacted with a flat hammer, and then coated three times with cement material consisting of a lime-clay mixture. The proportion of lime to clay is 1 : 3 for the first coating, 1 : 1 for the second, 3: 1 for the third. The thickness of each coating is about 1 to 1.2 cm. If the wall of the well is of hard soil, application of two coatings may be enough.

In regions where lime is unavailable, red clay is used after first being sieved to separate stones, then submerged under water for at least three days, and repeatedly stirred until it becomes a uniform thick slurry. On the wall of the well, holes with a diameter of 7 to 10 cm and a depth of 17 cm are dug at a spacing of 15 cm and are staggered in arrangement. On the third day after digging, the holes are then filled tightly with balls of red clay. The area between the holes is then coated over with red clay to a thickness of about 2 cm at the upper part of the well, 3 cm in the middle part, 4 cm at the lower part and 5 to 6 cm on the bottom. As soon as the clay slurry is coated, it should be trowelled smooth. Subsequently, the wall has to be hammered once or twice a day, and repeated about 30 times, when the impervious layer appears shiny. The dry density will then reach 1.78 t m^{-3}. A protection layer of sand and gravel is then provided on the bottom of the well, and a platform is laid with bricks or stones at the mouth of the well. A cover is added to prevent pollution and maintain a certain humidity in the well for protecting the impervious layer.

A water cave is another type of water-storage well. Its chamber is similar in shape to the old cave dwelling in this region and so it is referred to as a water cave. The water cave holds much more water than the water-storage well. A cross-section is shown in figure 11-3. It is usually constructed on the terrace or near to any earth precipice with a height of more than 3 m. Its construction is begun by digging an opening (1.6 to 2.0 m × 1.0 to 1.2 m) in the foot of the terrace as an entrance to the water cave. When the entrance corridor reaches 2 m deep horizontally, excavation then proceeds laterally left, right and downward. An inflow canal lined with masonry is then built for diverting rainwater through the cave entrance to the water cave. An intake pipe is inserted extending out of the cave wall about 5 cm, and a stone protection layer is placed on the bottom of the cave. Near the entrance, a receiving pit is dug to the same depth as the water cave. Along the bottom of this pit an outlet hole is excavated in which a tile (or earthenware) release pipe is laid from the water cave for releasing the stored water. The pipe end should be 15 cm higher than the bottom of the water cave. This release pipe is equipped with a gate which can be operated from above the entrance. The mouth of the water cave is then excavated and a platform of bricks or stones is laid. The impervious wall layer is made in the same manner as the water-storage wells.

After completion of a water-storage well, it is necessary, even before storing rainwater, to ensure that a certain amount of water is in the well so as to maintain relative humidity. This is to prevent the impervious layer from cracking due to drying and to ensure its proper function. Also, during the water storage, the water level in the well should not rise to the delivery section, and the mouth of water-storage wells should be kept covered to ensure the cleanliness of the water. The well is usually cleared of silt deposits about once every 3 to 5 years.

The water-storage well is adaptable to local conditions. It can be constructed either on the slope of high land or on the flatland areas where water flow can be collected through any gully. Rainwater accumulated in the well has a high rate of utilization, because evaporation and seepage losses are reduced to a minimum. The construction technique is simple and can be easily stered by most people. Moreover, the well uses only local materials, and so the cost is low.

Generally a water-storage well, having a capacity of 15 m^3, needs only 25 to 30 workdays and 100 to 150 kg of lime. During the construction of water-storage wells, attention must be paid to the following points:

(a) local geological defects may cause construction problems and will result in poor quality of completed wells;
(b) improving the construction method and raising the productivity of the methods and;
(c) improving the catchment for collecting rainwater and improving water quality.

Ponds

Ponds are used as water-storing facilities in the Loess Plateau Region. They are dug in depressions surrounded by low levees to store the surface runoff and the water from mountain springs and creeks. Storage capacities range from several hundred to several thousand m^3.

Ponds are usually constructed near villages, roads, slopes, lowlands or the heads of gullies, where surface water may accumulate to the amount required. The soil at the site of the pond must be fairly stiff, preferably a clay. As the pond is surrounded by levees, it is constructed partly by excavating and partly by filling. The levee usually has a top width of about 1 m and side slopes of 1 : 1.5. Its top should be higher than the maximum water level by 0.5 to 0.7 m. The water depth in the pond should be between 2 to 3 m. The levee must be of good quality construction to ensure safe storage of rainwater. In order to prevent overflow and erosion of the levees, a spillway must be installed, which is generally lined with turf or bricks.

The bottom of the pond must be carefully treated to prevent seepage as follows:

(a) along the bottom of the pond a layer of red clay with a thickness of 0.1 to 0.2 m is firmly compacted;
(b) if loess soil is used on the pond bottom it must be tamped, and then 200 to 300 g of salt must be spread for every square metre. After another tamping, this salt will form an impervious layer (as in treated ground catchments—chapter 5);
(c) in place of (a) or (b) a layer of red clay-sand-gravel mixture (proportion 6:2:2) can be used. This is placed on the pond bottom with a uniform thickness of 0.2 m and then tamped heavily. This will also satisfy the requirement for seepage control;
(d) where available a layer of lime-clay mixture (proportion 1:5) will also result in a good impervious layer.

Another pipe equipped with a wooden plug or a small valve is buried beneath the levee during construction to control the flow out of the pond. When the water level is lower than the surrounding fields, it is necessary to pump water if it is used for irrigation. And in order to conserve the water it is preferable to plant poplars or willows around the edge in order to reduce evaporation. The pond should also not be allowed to dry up otherwise the impervious layer will be damaged.

The pond makes use of lowland runoff and it is simple in technology and construction. It requires a little work, but has a relatively large storage capacity. However, the disadvantages of ponds are their high evaporation losses and low water quality in comparison with water-storage wells. Hence, the pond water is usually used for irrigation and stock water supply.

Ground Water Storage

In the loess highlands in China there are many small rivers which dry out each year. These are referred to as 'gullies'. Impervious cut-offs can be built through the deposits in the channels of such gullies, or even into ravines or rivers to intercept the ground water underflow. These are called 'underflow interception projects' in China, but they are similar to the concept of the 'qanats' of the Middle East, the interception galleries in Europe, and percolation tanks of India (chapter 5).

A project of this type at Guishi Gully, Pingding County, Shanxi Province, generates a discharge of of up to 200 ls^{-1}, solving the problem of drinking water for 60,000 people! The sites for these underflow interception projects obviously must be selected with care. The best places are where there are long gullies with large catchment areas inside narrow rock outcrops on both abutments; and generally, where river bed deposits are shallow, preferably 5 to 10 m deep.

Underflow interception projects comprise three parts, the impervious cut-off, the collecting passage (gallery) and the conduit. There are two types of cutoff: the central core (figure 11-4) and the sloping core (figure 11-5). Both serve the function of intercepting flow. The cutoff is mostly built up out of clay, stone masonry, or concrete. It should, in general, extend to the bedrock or to the impervious layers. The base thickness of the clay core is 1/3 to 1/4 the height of the core and the top thickness is 1 to 1.5 m. Special attention should be paid to the construction quality of the contact between core and bedrock and abutments.

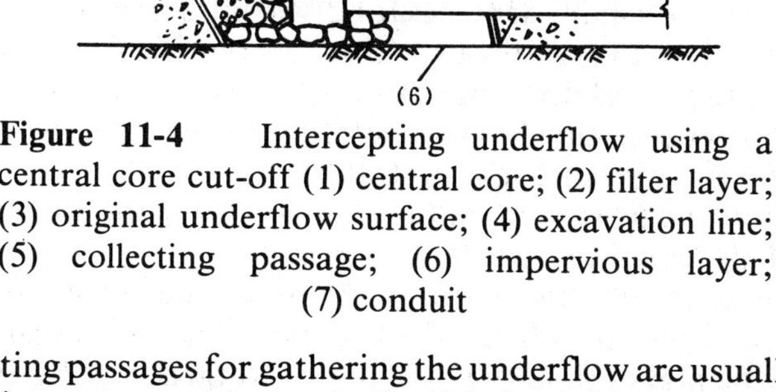

Figure 11-4 Intercepting underflow using a central core cut-off (1) central core; (2) filter layer; (3) original underflow surface; (4) excavation line; (5) collecting passage; (6) impervious layer; (7) conduit

The collecting passages for gathering the underflow are usually rectangular in section. Their upstream sides are built with dry-laid masonry or partly with masonry laid with cement mortar. Their downstream sides are wholly masonry laid with cement mortar. For the convenience of construction and management, the width of the passages should not be less than 80 cm, and their heights not less than 1 m. Filter materials should be backfilled against the outside faces to ensure safe seepage.

Diversion conduits of concrete pipe or stone masonry culvert lead the underflow gathered in the collecting passages out to the canal downstream. Inspection wells are provided at certain intervals along the diversion conduits, in the same fashion as qanats (chapter 5).

Underflow interception projects can solve the problem of lack of surface water, because the flow is steady and the quality of water is good. However, such projects demand good geological conditions and a large amount of construction.

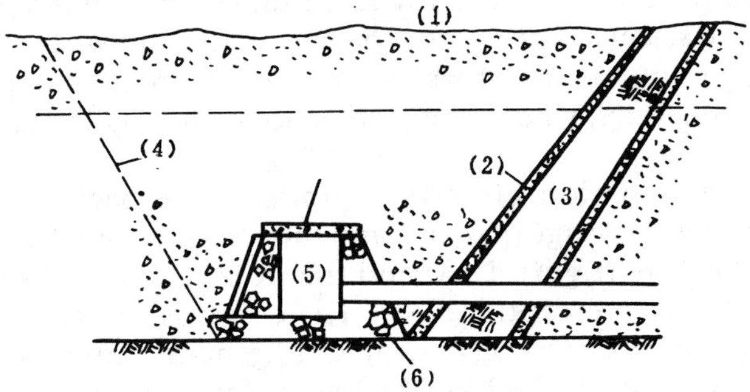

Figure 11-5 Intercepting underflow by sloping core cut-off. (1) original river bed surface; (2) filter material; (3) sloping core; (4) excavation line; (5) collecting passage; (6) impervious layer

STORMWATER HARVEST—WARPING

In China there are many schemes for diverting storm and floodwater from gullies and also diverting rainwater from barren hillsides into villages and farmland. The purpose of diverting this water is to fertilize the soil with the silt and organic matter contained in the water and simultaneously to irrigate the crops. This is an effective measure for fully utilizing water and soil resources in the Loess Region and has the advantage of decreasing the menace of water and soil erosion. This technique is called 'warping'.

According to recent observations in China, one hectare of farmland can generally retain 1,995 to 3,000 m^3 of water. The soil moisture of warped farmland may be higher than that of unwarped farmland by 10 per cent, and even by 77 per cent in drought periods. The organic matter increases by 21.8 per cent and nitrogen by 13 per cent on warped land, resulting in a marked increase in agricultural production.

Diverting the flood for warping decreases flood hazards through dispersing the runoff and cutting down the flood peak. One example is warping in the district along the Bali River in Dingbian County, Shanxi Province. Its area is approximately 50 km², with a total annual runoff of 23.4 million m³ and a sediment yield of 3.70 million m³ which is completely stored by diverting the flood for warping. Another example in the same province can be seen at the Mawan Creek in Jingbian County. The catchment area here is 101.2 km². In nearly one hundred years (except in one year of extreme flood) no water and silt have been released from the gully since the practice of diverting the flood for warping began.

Warping can also turn a barren gully or floodplain area into fertile land with a consequent expansion of cropland. In some localities, floodwater is diverted for leaching saline soil or for ameliorating saline-alkali land (as in India, chapter 10).

The Warping Process

Warping may be accomplished in several ways depending on the source of flood. In the first case a dam can be constructed at the mouth of a gully to divert muddy water on to farmland at the foot of the hill and on to the flats on either bank of the stream (figure 11-6). A second type is when the floodwater is diverted from a large or medium river. Here warping is mostly done on the river flats or floodplain. A third technique is the use of intercepting ditches and diversion canals. These are constructed along contour lines to intercept overland flow on hill slopes and the water is used for warping river flats and terraced fields (figure 11-7). The fourth method is the collection of overland flow from village areas and roads. This technique makes use of courtyard or cleared open areas for harvesting muddy water which can then be used directly in warping.

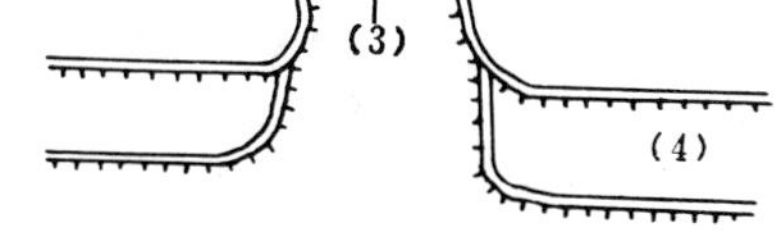

Figure 11-6 Warping by gully flood. (1) gully; (2) diversion canal; (3) earth dam; (4) terraced field

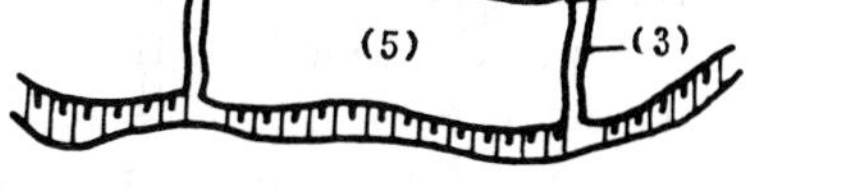

Figure 11-7 Intercepting flow on hill slopes. (1) barren hillside; (2) contour line; (3) diversion canal; (4) intercepting ditch; (5) terraced field

The depth of warping varies with the type of crops and their growing period. The depth of warping for some crops in the north part of Shanxi Province is shown in table 11-2.

Farm projects are the key for ensuring effective warping. Generally, the farm plots are arranged along contour lines. The borders of the plots are bounded by small levees of earth or rock, with the size of the plot depending on the slope of the land and the pattern of cultivation. The plot area is usually about 4 to 10 ha for flat land and 0.3 to 7 ha for hilly loess land.

The levees also play an important role in storing silt and water. They must be finished before the flood season, and be of suitable height (0.3 to 0.5 m, and as high as 0.5 to 1.0 m on steep slopes), and width (0.3 to 0.5 m).

Outlets should be provided on field plots for de-watering any excess water. This will ensure the safety of levees and facilitate warping. During the process of warping the colour of the water is important. When the muddy water becomes dark yellow, it then contains a greater amount of organic matter and has greater fertility. In China floodwater which contains red clay or has higher alkalinity is considered unsuitable for warping. Warping is the sort of work that is usually performed during some critical point in the season when climatic conditions are not very good, therefore it calls for good organization.

Examples of Warping

The Zhaolao Gully warping district is situated in Fuping County, Shanxi Province. Its history can be traced back more than 2,300 years to the time of Qin Kingdom of the Spring and Autumn Period. It diverts the flood from Zhaolao Gully, upstream of the Shunyang River, to irrigate 2,260 ha of farmland located at the alluvial fan.

The length of the main gully is 24.2 km. The catchment area is about 200 km^2. The soils in the catchment are mainly loess, with an annual precipitation of 600 to 700 mm. Generally, there are about 10 flash floods a year, occurring mostly in the period from the second decade of July to the first decade of August. The peak discharge is 120 to 160 m^3s^{-1} with a maximum of 327 m^3s^{-1} in extraordinary cases. The organic matter content of the muddy water is 1.4 to 2.02 per cent, whereas the nitrogen content varies from 0.062 to 0.107 per cent.

Along the gully banks of 9.5 km long in the flood irrigation district, there are 37 diversion inlets and 24 diversion canals. The total diversion capacity is 205 m^3s^{-1}. An overall view of the district is shown in figure 11-8.

The effectiveness of warping in this scheme is remarkable. The water content of the soil is notably increased. The top 30 cm soil layer of warped land has a water content 2.8 to 4.4 per cent higher than that of unwarped farmland. After warping, the soil composition also changed. The sand content decreased from 71.2 per cent to 46.9 per cent, and the clay content increased from 28.8 per cent to 53.1 per cent. The soil is thus ameliorated.

According to the calculation that 2,700 m^3 of muddy water must be

Table 11-2 Depth of warping for some crops in the north part of Shanxi Province

Kind of crops	Sowing period	Harvesting period	Warping period and depth				
			May	June	July	August	September
Hemp	The last decade of March	The second decade of September		Shallow	Medium	Deep	Shallow
Spring wheat	The first decade of April	The second decade of July				Deep	Deep
Pea	The first decade of April	The second decade of July				Deep	Deep
Millet	The last decade of April	The second decade of September			Medium	Shallow	
Black beans	The first decade of May	The last decade of September			Shallow		
Broom Corn, Millet	The first decade of June	The first decade of September	Shallow		Medium		
Buck-wheat	The last decade of June	The second decade of September	Deep	Shallow			
Winter wheat	The first decade of September	The last decade of July					

Note: Shallow—about 0.15 m; Medium—0.25 m; Deep—less than 0.5 m

Figure 11-8 Plan of Zhaolao Gully flood irrigation district. (1) Zhaolao Gully; (2) Jiangjia Gully; (3) Lanshan streamlet; (4) seasonal streamlet; (5) old irrigation district; (6) ancient irrigation district; (7) new irrigation district; (8) Hongwei irrigation canal

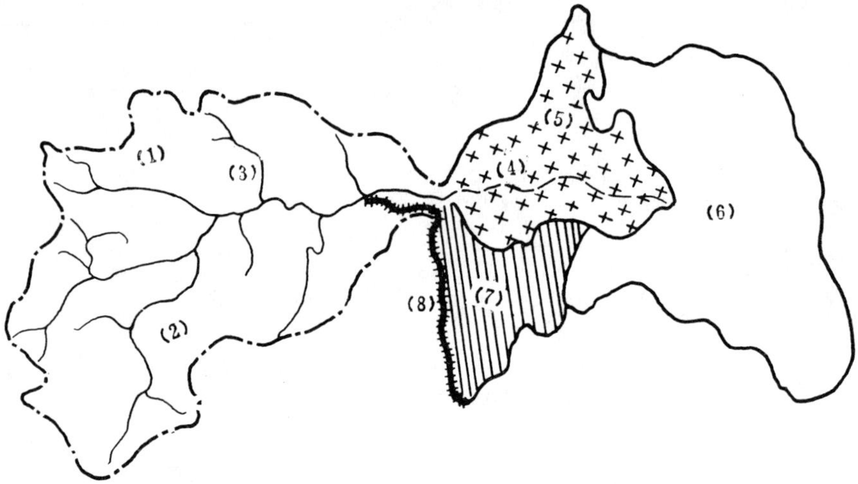

diverted for each ha of farmland, the nutrients that are carried by the water on to one hectare of farmland will be as follows: 560 kg nitrogen; 14 kg effective phosphorus; and 60 kg of effective potassium. A view of warped farmland is shown in figure 11-9 where the new silt layer can be seen cracking just prior to being cultivated. A comparison of crop yields on such soil is shown in table 11-3.

Table 11-3 Comparison of crop yields on warped land

	Maize ($kg\ ha^{-1}$)	Wheat ($kg\ ha^{-1}$)	Millet ($kg\ ha^{-1}$)
Warped farmland	4500	4129	3829
Unwarped farmland	2252	900	1350
Increased yield	2248	3229	2479

Because floodwater is now diverted totally for warping over many years in the Zhaolao Gully, very little is discharged downstream. The flat areas along the Shunyang River downstream have been reclaimed gradually. They extend along a length of 40 km downstream of the mouth of the Zhaolao Gully and have all been turned into fertile land. The riverbed has since disappeared. This river with a total length of 60 km has thus become the first river in the country that carries no silt into any major watercourse.

Figure 11-9 Warped farmland in Zhaolao Gully

STORMWATER HARVEST—SILT TRAPS

Silt traps, constructed of earth or rock across a gully, will impound floodwater and catch silt. Because of the silt accumulation small gullies are transformed into farmland, a process similar to the silt traps used in southeastern USA and Central America (chapter 8). This technique has the same effect as that of diverting the flood for warping, that is, fertile land is built up at the expense of top-soil erosion.

In recent years silt traps are being constructed in groups, along with unified planning of stream basins. They constitute a multiple-purpose system for agricultural production, silt and flood detention and irrigation.

Silt traps in China are built of earth or rock with a spillway and an outlet. Because the principal purpose of the trap is to catch silt for forming farmland, supplemented with water storage for irrigation, the requirements for design are somewhat different from those for ordinary dams. The geological and operational conditions required by silt traps are much simpler and less rigorous.

The location of silt traps should be at a place where a large storage can be obtained and where the slope of the gully is fairly gentle. Most silt traps are homogeneous earth dams. In any area where soil materials are unavailable in sufficient amount, rockfill or earth-rock construction can be used.

The height of the trap is determined according to the flood discharge, and volume and total amount of silt to be trapped (figure 11-10). In the north part of Shanxi Province, a 10-yr (10 per cent) flood is used as a model for designing a silt trap having a height less than 10 m and controlling a catchment area less than 2 km^2. Under certain conditions, a 20-yr (5 per cent) flood is adopted.

The design dam height is determined as follows:

Design dam height = Height of siltation + Depth of flood detention + Freeboard.

The height of siltation is determined according to the storage capacity required for holding 3 to 5 times the average amount of silt where:

$$\text{Annual silt volume} = \frac{(\text{Average soil loss per unit area}) \times (\text{Catchment area})}{\text{Dry density of silt}}$$

The depth of flood retention is equal to the water-depth above the spillway crest. In general this is 0.5 to 1.0 m and at most 1.5 m, and freeboard is 0.5 to 1.0 m.

The silt trap section is dependent on dam height, soil properties, construction method and traffic requirements. The slope is 1 : 1 to 1:3 and the crest width is 1.5 to 6.0 m (figure 11-10).

A spillway has great influence on the cost and safety of the project. It is composed of an open channel chute and stilling pool (figure 11-10). Once muddy water has entered and the silt has settled the clear water must be released from an outlet. The release of clear water from the outlet allows space to store more floodwater. The water released is used for irrigation or for warping if it is excess floodwater. The outlet consists of an inclined intake pipe, conduit section and stilling pool. The design principles and structure of the spillway and outlet are the same as those in the case of ordinary dam spillways and outlets.

The silt deposited behind the silt trap will eventually form cultivable land. However, cultivation on such land will hamper the passing of floodwater and could endanger the dam during the flood season. This contradiction between cultivation and flood prevention is resolved by using any of the following five methods of farming.

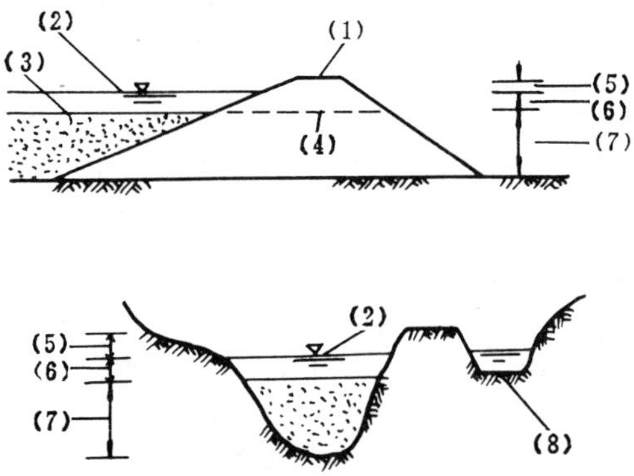

Figure 11-10 Cross section of a silt trap and channel across a gully. (1) crest; (2) highest water level; (3) elevation of silt deposit; (4) elevation of spillway; (5) freeboard; (6) depth of flood detention; (7) height of siltation; (8) spillway

Flood Storage on the Upstream Side and Land Cultivation on the Downstream Side: This technique is illustrated in figure 11-11. It is of use especially in a gully with small floods. Here a number of silt traps are built in stages downstream. After the silting up of the first reservoir downstream, the second trap is built on the upstream side. Similarly, a third and others are built further upstream. The upstream silt trap is mainly used for flood detention and can be used side-by-side with cultivation of deposited land. The downstream silt trap is mainly used for farming, which develops simultaneously with silt deposition.

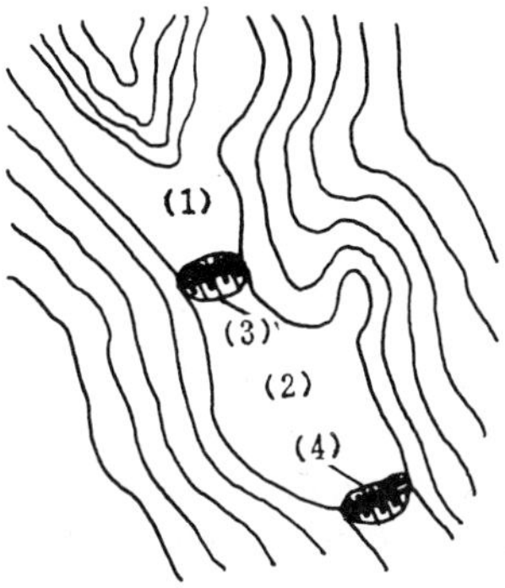

Figure 11-11 Flood storage on the upstream and land cultivation on the downstream side. (1) flood detention; (2) cultivation; (3) downstream dam; (4) upstream dam

Flood Storage and Farming in Rotation: After a series of silt traps have been built in a gully, most of them are used for cultivation, while a few are used for flood storage. When these storage traps are silted up, some of the traps used for farming are raised by dam heightening to function in flood storage again.

Combination of Flood Release and Flood Diversion for Warping: If a series of traps has been built in a gully and if it is impossible to raise the dam height or to build another trap for flood storage, the flood water may be diverted away and used for warping along both banks so that the silt can be deposited evenly (figure 11-12).

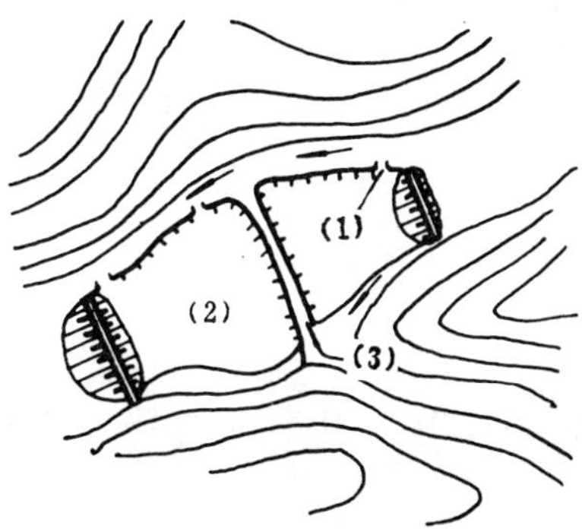

Figure 11-12 Flood release as the main feature, combined with warping. (1) inlet, (2) land inundation; (3) flood-way

Partial Release and Partial Warping: After a series of silt traps has been built in a gully and inflow can be dispersed, a system of partial release and partial warping is adopted, so that most of the floodwater is diverted for warping, and the rest is released out of the gully.

Provision for Storing Muddy Water and Releasing Clear Water: After a gully is well under control and the amount of floodwater is not large, the silt traps can be provided with outlets which will release only the clear water so that the sediment is retained.

STORMWATER HARVEST—CASE STUDY

Jiuyuan Gully, a small tributary in the middle reaches of the Wuding River, is situated in Suide County, Shanxi Province. It has a catchment area of 70.1 km^2, with a main channel 18 km long. Hence the mean annual rainfall is 516.9 mm and in the basin there are 9,334 inhabitants with 2,130 ha of cultivated area. The basin is under the jurisdiction of Juiyuangou People's Commune and the people there are all engaged in agricultural production.

In the basin soil erosion was a serious problem, especially because the gully density was 5.34 km per km^2. Before its control, the annual amount of erosion was 1,270,000 metric tonnes, corresponding to a soil loss of 18,120 tonnes km^{-2} per year.

In this basin comprehensive control has now been carried out with due attention to topographic and landform features. The two chief measures adopted were water harvesting techniques as follows:

Contour Farming

On sloping land terraced fields were built bounded with earth dykes. Rotation cultivation of grasses and crops was also implemented, in addition to shortening the length of slope and deep ploughing. Amelioration of the soil was the result.

Silt Traps

Throughout the gully, silt traps were built from the upper reaches to the lower reaches to check the runoff and silt. All the gully slopes are afforested in order to increase the area of vegetation.

Up to 1974, 727 ha of terraced fields and 311 silt traps were built throughout the whole basin. In addition, 30 small reservoirs were constructed with a total storage capacity of 1,118,000 m^3. The irrigated area had reached 170 ha which is 11 times more than that before control. These control features in the Jiuyuan Gully basin are shown in figure 11-13. After flood control was

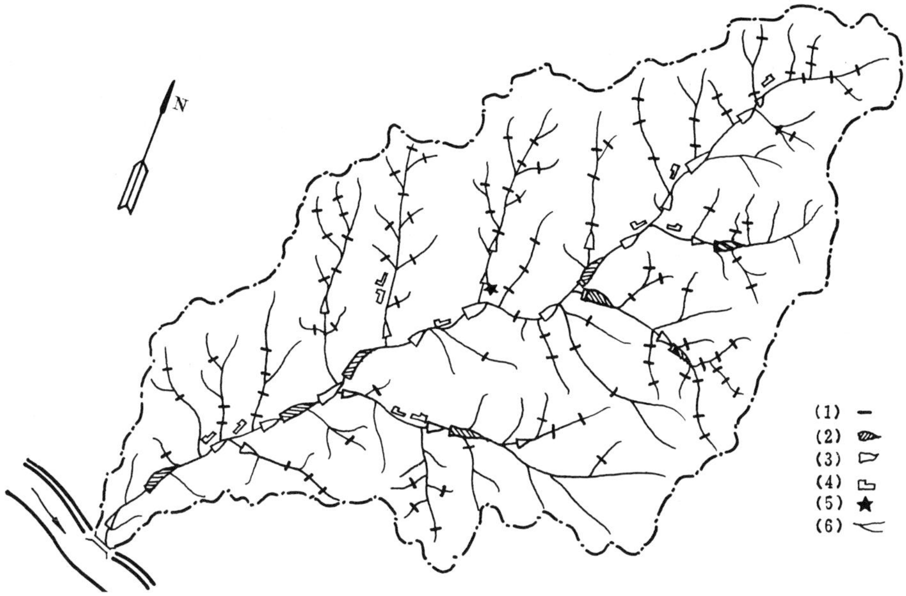

Figure 11-13 Layout of dams in Jiuyuan Gully basin. (1) small silt-arrest dam; (2) medium-size silt-arrest dam; (3) reservoir built by the commune; (4) village; (5) people's commune; (6) gully

Figure 11-14 After the control of the Jiuyuan Gully basin, flood is stored in the reservoir

achieved in the basin the amount of the average annual soil loss had decreased by 770,100 tons. The total amount of silt annually trapped (58.5 per cent) is now slightly more than one half. On July 5, 1964, heavy rain of 132 mm fell in the basin, but the flood peak in the stream channel was 90 per cent less than that before basin control, and 63 per cent of the sediment was trapped. Thus, flood control has been achieved in the gully basin using water harvesting techniques. This has greatly improved farming and living conditions (figure 11-14) as is shown dramatically by the total grain yield in 1974, which was 2.3 times more than that in 1952. Water harvesting is truly a large success in China and will certainly contribute directly to the welfare of the Chinese people for many years to come.

CHAPTER 12

Harvesting Water from Dew, Mist and Snow

IT IS PERHAPS because most people in well-watered and hospitable climates are used to having water available in terms of tens of litres at a time rather than a few cupfuls, that they do not realize how welcome the reliable provision of even half a litre of water a day can be to an inhabitant of a dry area.

In arid and semi-arid areas, rivers and lakes are few and far between, and other sources of water require a combination of technical knowledge, money and incentive to make them work. What are the alternatives? This chapter looks at the less well known sources of water and the methods which could be used by those living in arid areas to exploit them and we are indebted to Jones (1979) for most of the ideas.

It is rarely considered that there is a large natural reservoir of water in the world in the form of water vapour. For example, the typical relative humidity at Wajir in northeastern Kenya at 09.00 hours is 75 per cent and the average dry bulb temperature at the same time is 25°C. At that instant in time, at an atmospheric pressure of 990 millibars, the air over Wajir contains about 15 grams of water per kg of air, or nearly 18 grams per m^3. Assuming that the kilometre of air near the ground is well-mixed and contains the same proportion of water throughout, in one cubic kilometre of air over Wajir there is about 16 million litres of water. The assumptions involved are generous, but it gives an idea of the potential of such a source, and the figures are typical of Kenya's arid areas.

The problem is how to realize this potential. One could not expect to remove more than a fraction of one per cent of this water, but even that would yield something like a thousand litres per square kilometre per day.

It should not be forgotten that the soil which appears to be dry contains large amounts of water within the soil pores. Monteith (1957) gives a diagram showing that a dry sand, no longer able to support vegetation, nevertheless still contains 2.5 per cent by weight of water, equivalent to about 50 litres in a cubic metre of 'dry' soil. The hygroscopic properties of some soils are well known, and as an example the moisture content in the surface layer of an Indian Black Cotton soil increased from 2.6 to 7.7 per cent of dry weight by absorption overweight. Unfortunately, a soil dry enough to absorb moisture from the atmosphere is rarely wet enough to yield it to plant roots. However, Halacy (1966) describes experiments conducted in Japan and frequently repeated elsewhere, in which plastic sheets or miniature glass houses have been used to distil this water out of the ground and condense it in a container. About a litre of water a day can be collected from a square metre of ground in this way. The principle of operation is similar to the solar still, but one does not need even a brackish supply of water as a raw material. The soil moisture is replenished constantly by diffusion from deeper layers of soil and by absorption of atmospheric moisture.

Soil is not the only natural material capable of giving up water by evaporation. A case in point is seen in Australia where many overland travellers and pastoral-industry workers in the arid zones carry with them the basic elements of an emergency survival kit which includes a suitable container for water and a sheet of transparent plastic material. If caught without potable water, the user can excavate a hole in the ground, at the bottom of which is placed the open container surrounded by freshly cut leaves and grass. The hole is then covered by the plastic sheet which is weighted down by a stone to form a cone whose vertex is just above the container. When exposed to solar radiation, the evaporated water vapour from the plant material condenses against the cooler, transparent plastic sheeting and in due course, results in a modest dripping of water into the container. This useful, if simple, technology, is the basis of larger-scale solar desalination plants developed by the Australian Commonwealth Scientific and Industrial Research Organization for use in small settlements which need to rely on saline water.

A great deal has been written about dew as a source of water. Review papers or brackish by Milthorpe (1960) and Stone (1957) show the degree of controversy that exists between experts, many of whom express diametrically opposed views.

Enthusiastic, and rather subjective investigations by Hubbard and Hubbard (1905) asserted that 'There is still in England at least one wandering gang of men . . . who will construct for the modern farmer a (dew) pond which, in any situation, will always contain water, more in the heat of summer than during the winter rains'. However, Martin's (1910) three year investigation into dew-ponds gave almost conclusive proof that the ponds were replenished more by mist and fog than by dew. Many writers have since indicated that the maximum theoretical deposit of dew in a single 10-hour night, computed from energy considerations, is of the order of 0.8 mm, certainly not enough to replenish a 'dew-pond'.

More recent investigations into dew have verified this maximum figure experimentally. The measurement of dew is difficult, and artificial dew gauges so far devised cannot properly imitate the physical characteristics of natural collectors such as plants and rock. Also, it is difficult to distinguish between dew formed from truly atmospheric moisture, dew formed by distillation of soil moisture, and guttation, the moisture exuded by plants from their leaf tips.

There are many cited examples, though few quantitative measurements, of plants being grown by reliance on dew. Went (1955) suggested that in California, tomatoes can be grown in an area where no rain falls during the growing season. On Lanzarote in the Canary Isles, there are small depressions a metre or so across, lined with black volcanic pebbles said to condense dew which then trickles to a solitary onion plant in the middle. Grape vines are said to suffer less from drought than normal when grown on rocky slopes; near Amman in Jordan, tomatoes are said to be grown almost entirely on dew, and it has been claimed that the spreading thorn tree, by its shade, derives enough moisture to survive from the atmosphere. White (1906) mentions that some dew-ponds are overhung with 'two moderate beeches that doubtless afford it much supply', Milne and Milne (1964) describe how, on a clear starry night, they stood under a tropical rain-tree in Trinidad and became soaked by the drips from the leaves. Whether this last experience was the result of condensation dew or guttation they were unable to tell, but at least it gave rise to a thick carpet of grass under the tree, whereas in the exposed area away from the tree the grass was thin and dry.

Arvidsson and Hellstrom (1955) working in Sweden and Egypt have indicated that dew could provide water in usable amounts. They constructed large funnels of various materials, supported about 1 m above the ground. The funnels collected dew on about a third of the nights on which they were exposed. (Cloud cover prevented radiation and the subsequent cooling necessary for dew formation on the other nights.) Dew which collected on the inside and outside of a wooden funnel with a plan area of one square metre gave a maximum collection on one night of 0.4 litre.

Meteorological data for many parts of the arid world are sparse and give little indication as to when and where dew can be expected to form. However, observant persons will have noticed the row of drips frequently found under the eaves of buildings in many arid parts of a country such as Kenya in the early morning, even in warm, dry weather. A medium grade house with a plan roof area of 29 m^2, would, applying the above average figures, collect 320 litres of dew water a year. This may seem a small amount compared with the amount of rainwater which could be normally collected (chapter 3), but this amount could be significant if collected at the end of a long, dry period. It is surprising to note how many buildings in sub-humid Africa, and even in dry areas, do not have gutters and/or storage tanks.

Monteith (1957) conducted a series of careful experiments on dew on a playing field in England. Though the environment in which his experiments were conducted was very different from that in arid regions, he did verify one that is often mentioned, that is, the necessity of a light wind. In order to condense the moisture present in the atmosphere as dew, it must be brought into contact with the condensing surface; a light wind speed, i.e., 2-3 $m.s^{-1}$ will stir up the 100 metres or more of air nearest the ground and bring most of it into contact with the condensing surface. A stronger wind, over 5 $m.s^{-1}$ always resulted in net evaporation. Anon. (1961) and Deacon *et al.,* (1958) have reviewed the possibility of choosing areas with significant amounts of dew for planting crops which can use it effectively and thus enhance dew formation in places where there is none at present. Went (1955) has suggested that rows of trees fact should be planted in suitable locations to reduce wind speed to that required for dew formation. Though we know that the maximum dew precipitation is small, various writers have referred to the capability of plants to 'attract' dew. It is suggested by Slatyer and McIlroy (1961) that plants might concentrate the precipitation of dew, but only at the expense of surrounding areas. Certainly, if an artificial dew collector could do this as effectively as the tropical rain trees in Africa, then there may well be scope for developing this source. A vast amount of data collection and investigation on dew has been done by Duvdenani *et al.*, (1957) in Israel with particular emphasis on the part it plays in arid land agriculture. They are convinced that dew absorbed through the leaves of plants plays an important part in maintaining plant function during periods of drought. Gindell (1973) for example, described how pine, tamarisk and eucalyptus seedlings were planted in the Judean Hills. Near each plant a plastic sheet was arranged, 1.3 m^2 in area. The dew collected on the sheet was led to the plant pit. Saplings planted even during hot or dry conditions, normally unsatisfactory for survival, nevertheless survived until the rains came four months later. Those without the dew supply dried up and died within a month. The plastic sheet may also have had a significant effect on preserving soil moisture from evaporation.

Dew has often been credited with providing large quantities of water which have actually been the result of interception of mist and fog. In the case of mist and fog, the moisture is already condensed, but is in the form of small droplets floating about in the air, too small to fall to the ground. Martin (1910) acknowledges the role of mist interception as a source for his dew-ponds, and it is likely that some of the instances of plant survival mentioned above owe more to mist or fog interception than to direct condensation of dew.

The significance of mist and fog interception was noted by Hursh and Pereira (1970) when considering the water economy of the Shimba Hills, in Kenya. They noted the significance of mist and fog interception and have related this to forest formations or belts of trees which have been planted for the specific purpose of straining moisture out of the fog. Similar work in Japan was carried out by Hori (1953) and Kashiyama (1956).

The process of mist interception by trees has been used for many years by the people inhabiting an escarpment near Kitui in Kenya. This area faces the moist southeasterly winds. Here a tree is chosen, such as a gum tree, which has its branches pointing upwards away from the stem. A rope is coiled around the stem with its loose lower end resting in a bucket. This arrangement will collect water which originates from mist and fog interception in the upper branches. Some work on this subject was done at the experimental station at Muguga in Kenya (Anon., 1961). Fog-catching gauges located there in preliminary trials show up to three times as much water was caught in them as in a standard rain gauge.

Nagel (1956, 1962) investigated mist and fog interception on Table Mountain and along the coast of South West Africa. Table Mountain has an almost permanent covering of cloud, nicknamed the 'Tablecloth'. Nagel set up rain gauges, some of which had fog catchers mounted above them. The fog catcher consisted of a gauze cylinder with its axis set vertically above the axis of the rain gauge. His results showed fog interception over a year to be equivalent to 3,294 mm of rain, compared to actual rainfall over the same period of 1,940 mm. On theoretical grounds, he concluded that the amount of water available in the fog was about 12 times that which was actually intercepted. Further experiments gave similar results in other locations. One surprising result was that the amount collected by a flat gauze screen mounted on a lorry and driven through the fog at different speeds did not vary very much with the speed. The results of such experiments are often complicated by the fog catcher's ability to intercept rain, drizzle and dew as well.

Twomey (1957) in similar experiments in Tasmania, found that a section of flat mesh 1 m high set vertically above a 20 cm rain gauge yielded about ten times more water than a rain gauge without mesh. Taking the water content of cloud to be 1 gm m^{-3}, he points out that with a light wind of only 4 m s^{-1}, this represents a potential catch of 14 litres of water per hour passing horizontally through a square metre. Nagel quotes a measured wind speed of 13 km s^{-1} for Table Mountain, which could theoretically yield nearly 47 litres per hour. Twomey's mesh was of steel wire with 3.2 mm per aperture. Nagel's mesh was finer, with 1.6 mm per aperture and was similar to mosquito wire, a cheap and readily available material throughout the tropics.

A simple method of catching mist to promote the germination of seeds is to put a stake in the ground next to the seed. The stake will intercept mist and other precipitation, and guide it down to the seed. Kerfoot (1968) gives an extensive review of work on this topic which leaves little doubt as to its potential.

It seems that the ancients had similar ideas. Jumikis (1965) reported extensively on the discovery of rock mounds on hilltops in Theodosia (Crimea). These mounds, over two thousand years old, were 30 m in length, 25 m wide and 10 m high, and were connected by tile pipes to the coastal town. Appropriately known as aerial wells, they have counterparts in many places. Jumikis mentions the 'fogaras' of Tunisia, and the 'rotaras' of Morocco, and the 'teleilat' of Israel are common. Unfortunately, information on how they may have worked is hard to come by. Evenari, Shana and Tadmor (1971), after extensive investigations, concluded that the teleilat or stone mounds found on hillsides in the Negev desert (figure 12-1) had nothing to do with dew at all. Their observations failed to show any significant deposition of dew on the mounds, certainly not enough to cause the water to drip off and wet the ground underneath. Apparently the name *teleilat el einab* (translated from the Arabic as 'the hillocks for grape vines') is a misnomer based on wishful thinking, similar to the English dew-ponds which do not catch dew. They advance the theory that the mounds are the result of stone clearing from the rocky slopes, for the purpose of increasing runoff from the catchment (as discussed in chapter 4).

As for the Theodosian aerial wells, whether it was fog interception, dew formation, direct runoff or some hygroscopic quality of the rock resulting in what Jumikis and Anon. (1935) call 'dehydration of air', that was primarily responsible for the water that flowed down the tile pipes is a question yet to be

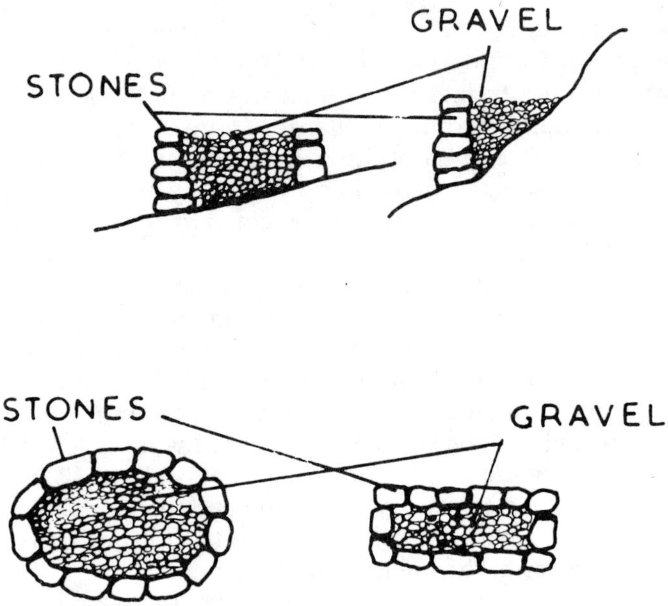

Figure 12-1 Mounds of stone filled with loose gravel in the Negev Desert

satisfactorily answered. However under certain favourable circumstances it does seem possible to harvest dew for agricultural purposes. One example from China illustrates an ingenious technique.

In Gansu Province, northwest China, there is poor annual precipitation, frequent wind, drought and a high rate of evaporation. The people here have had a long history of agricultural production and, with time, have developed a unique method of cultivation by means of which they grow delicious melons which are famous throughout the area. The melons are cultivated in a soil bed which is covered with a layer of gravel, 10 to 15 cm thick. The pieces of gravel range from 2 to 5 cm diameter, and in Chinese these farms are known as 'gravel fields for melon'.

This cultivation method depends on the following factors: First, the gravel layer maintains soil moisture, sharply minimizing evaporation losses; second, the gravel absorbs the sun's energy, thus raising soil temperature; third, in the evening, the gravel layer becomes cold, water condenses from the gravel and the soil forming drops on the surface of the gravel, and this moistens the soil, providing a satisfactory growing medium for the melons.

In the colder regions of the world, such as in Afghanistan, in Sheritala in Takhar Province, snow falling in the winter is stored underground, to be used as drinking water in the dry summer. The snow is collected in bags, loaded on

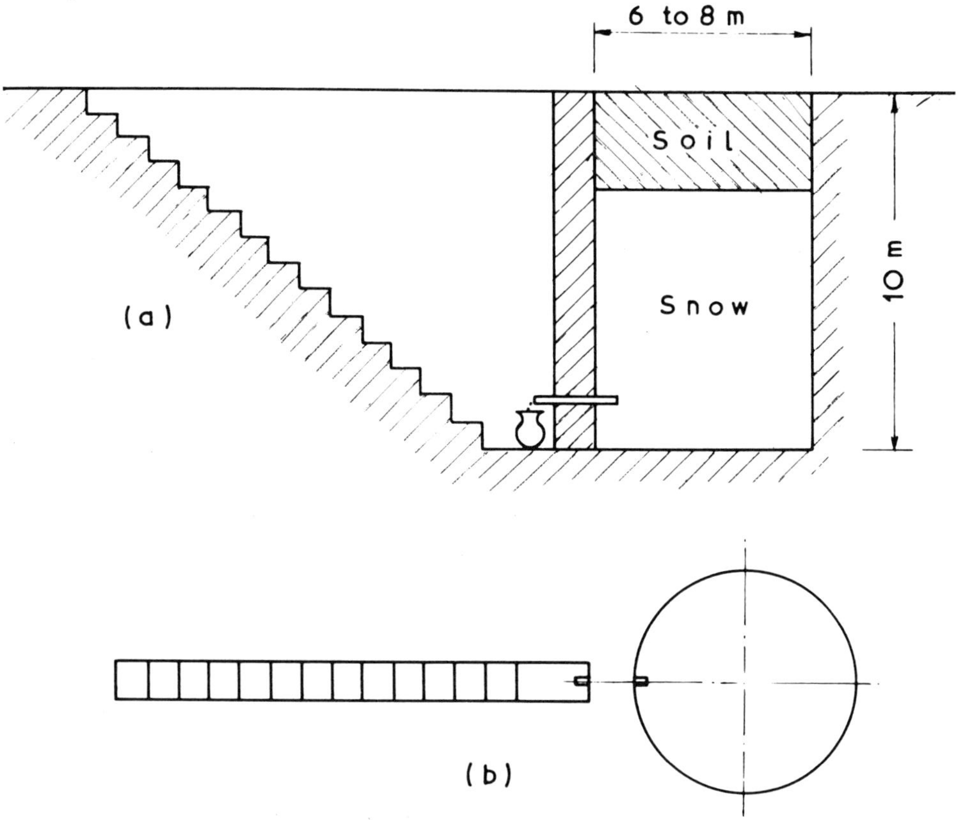

Figure 12-2 Snow storage for water harvest in Afghanistan. (a) cross section; (b) from above

donkeys and dumped in a pit in the village. The pit is 6 to 8 m in diameter and about 10 m deep. It is not lined, since the loess soil in this area (as in China, chapter 11) is easily compacted. The snow is compacted and the pit filled to within 2 to 3 m from the ground (figure 12-2). The remaining space is filled with earth, which acts as an insulator. A ramp or a series of steps is constructed from ground level down to the bottom of the pit. A small bamboo tube is driven through the wall of the pit into the snow some 50 cm above its bottom. As the snow melts slowly around the tube the water trickles down to a pot placed underneath. It is said that the storage of one such pit supplies the village (10 families) with its drinking water for two years! In this technique, losses due to evaporation and percolation are avoided. However, if snowfall is poor for two consecutive years, the entire village has to shift to a stream 40 km away in search of water.

CHAPTER 13

Design and Evaluation of Rainwater Harvesting Schemes

EVALUATION of any technology involves an assessment of its overall costs and benefits. This chapter will review the salient features of several technologies, providing us with a number of significant insights, and demonstrating that water harvesting techniques are an ingenious response to the problems and possibilities of subhumid environments.

DESIGN PROBLEMS

The common feature of any investigations connected with hydrological analysis of runoff control is that the daily discharge data of a natural catchment are observed directly or simulated and then are used as basic information to predict the amounts of water expected. The basic principles of rain harvesting are the same as those for runoff control. Likewise, in rainwater harvesting the natural distribution of the available water must be modified according to socio-economic demands. However, a large difference does exist in that the methods of hydrological investigation normally used in reference to a reservoir cannot be applied directly to the design of harvesting systems because of three factors:

(1) the collecting surface is usually not a natural system, thus observed data are not available to characterize the stochastic process and such information cannot be even deducted from other observations, since the runoff conditions are artificially modified;
(2) the extension of the collecting surface itself is a parameter to be determined, thus as in the Manda Island scheme (chapter 6) the design should concentrate on calculations of specific parameters characterizing the unit area of the artificially constructed collecting surface;
(3) the problem of long-term storage of water cannot be solved because of the limited storage capacity of rainwater harvesting schemes and, in the case of domestic supply, because of the deterioration of water quality unless it is treated.

The design of rainwater harvesting schemes must always start with the determination of the water demand to be met. In the case of domestic water supply the daily demand generally does not have much seasonal fluctuation and its value can be determined as the product of the number of inhabitants and the average daily consumption per capita. The water demand of plants should be considered when the systems are designed for agricultural use.

The extension of the catchment area by artificial treatment must not be calculated as the rate of the total demand and the yearly precipitation, for even if the latter is modified by considering the losses caused by interception and evaporation, in most cases the yearly storage still cannot be retained or stored by the limited capacity of most structures used in rain harvesting. In the case of agricultural utilization water consumption should be compared to the amount of precipitation of the growing season, or even the rate of demand. Available rainwater must be determined within short periods, when such rainwater is critical for the development of the crop. In the hydrological design of both agricultural and domestic use of rain harvesting, the critical length of consecutive dry days should always be investigated.

The basic time unit of hydrological design for harvesting schemes could be composed of the critical dry period and the rainy days preceding it. Then the comparison of the demand during this combined interval, to the amount of water available from the precipitation fallen in the first period of the interval, will result in a useful prediction of the necessary size of the collecting surface and the storage volume needed. By investigating a series of critical dry periods it is even possible to predict the existing demands.

A further design parameter will be the case for extremely high flow rates in the system, because the structures must ensure that excess runoff be passed by the system without causing any damage to the scheme or the local environment. After surveying the design problems the consecutive steps in the hydrological investigation can be listed in order to indicate the possibilities prior to the preparation of a pilot scheme or preliminary plan. The work that has to be done in advance is as follows:

(a) the mean annual depth of precipitation and its probability distribution in the investigated areas should be determined;
(b) water losses occuring in the collecting system which depend on local conditions should be investigated, e.g., on the degree of extension and/or treatment of the collecting surface, local climatic conditions, intensity of precipitation, temperature, and wind;
(c) graphs should be constructed to characterize the probable surface runoff from impermeable areas;
(d) the distribution of any extremely high daily precipitation expected should be determined and various probabilities worked out. This will be used as the design data for the calculation of the required maximum water transporting capacity of the structure;
(e) the relationship between the point value measured by rain gauges and the aerial average of the precipitation should be studied, because the latter decreases gradually as the area investigated increases. The

consideration of such reducing factors may be important, especially when the water is collected from larger catchments for agricultural utilization;

(f) a series of critical periods should be selected from a series of rainfall data. After consideration of the various corrections and considering the losses and the effect of general rainfall distribution, a series of expected specific runoff rates can be calculated from the unit area of the collecting surface. The possibility of this parameter can also be determined by investigating several critical periods by analysing the probable occurrence of critical conditions.

ECONOMIC CONSIDERATIONS

The construction of a rainwater harvesting scheme for a community water supply can be justified, as can any other investment, when all the benefits from such a scheme exceed all the costs. We know how to measure costs fairly well, although we are still learning how to measure costs which are external to water supply projects, such as the economic costs of ecological changes brought about in the construction and operation of such projects. As rainwater harvesting schemes are generally quite simple, we shall assume that all costs associated with them, both initial and recurrent, can be measured.

The measurement of benefits from a water supply is, however, more difficult. The economic benefit accruing to a consumer of any product is assumed to be the maximum amount that he would pay for it. Thus an individual consumer expresses his own estimate of the value of water by his personal demand curve, which has been explained clearly by one well-known economist (Samuelson, 1967) as follows:

> When water is very dear, I demand only enough of it to drink. Then when its price drops, I buy some to wash with. At still lower prices, I resort to still other uses; finally, when it is really very cheap, I water flowers and use it lavishly for any possible purpose.

Consequently, if the water supplied by a scheme is charged for on the basis of the amount consumed the benefit of the scheme can be conservatively estimated as the total revenue from the sale of water. Such a scheme would be economically justified if the present value of the expected revenues exceeds the value of the costs associated with it, where present values are determined using the appropriate discount rate. Sometimes, even in cases where benefits do *not* exceed project costs, a scheme might be justifiable, for example, in the case where a government is unwilling to allow its poorer citizens to die from thirst simply because they can't afford the price of water, or because there are additional benefits which are not revealed by the potential revenues from the sale of the water. These additional benefits could be due to several factors:

(1) A consumer's willingness to pay does not always measure all benefits which he might receive from the purchase of water. This is illustrated by a demand curve:

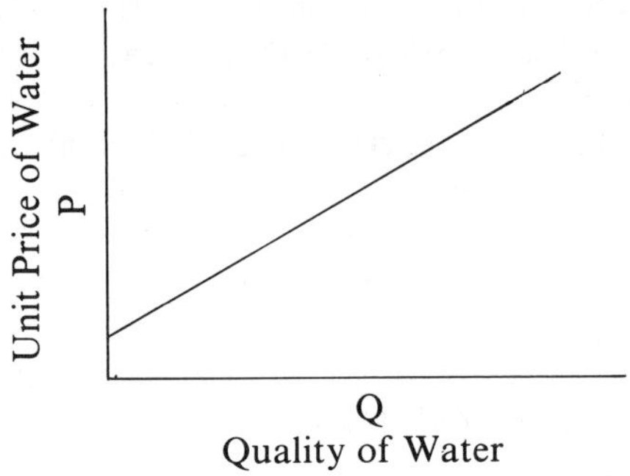

Thus, at unit price 'P' he will purchase a quantity 'Q' and the revenue is a partial measure of the value he places on the water. However, the *entire area* under the demand curve at the quantity Q is the full measure of the value of the water. The consumer would be willing to pay higher unit prices for the first Q units of water. No account is taken of the 'consumers' surplus', the triangular area under the demand curve and above price P, if the benefit is measured simply by his willingness to pay price P.

(2) People may not be as rational or well-informed as suggested by their personal demand curve. For example, they may not be in possession of all the relevant facts concerning the health benefits likely to accrue from using clean water. Their willingness to pay may therefore be an underestimate of the total value of water to them.

(3) There may be public health benefits which are external to those measured by an individual consumer's willingness to pay. In other words, the consumption of water by X may be of benefit to the health of his neighbour Y, but X may not take this into account in his own private market decision.

Because of these complications it is frequently not possible to measure all benefits which can be expected to result from a community water supply. The decision whether to invest in a water supply scheme can therefore seldom be made by the usual method of comparing costs and benefits. Instead the conventional approach is to assume that the provision of a supply of water can somehow be justified and that the problem is actually reduced to finding the least cost solution to supplying the estimated demand for water. Thus in the absence of a test for comparing benefits and costs the determination of the quantity of water to be supplied is crucial. Scarce resources will be wasted if a project is constructed to supply large quantities of fresh water for low value utilization by the consumers (e.g., garden watering). The selling of water at a price at least approaching the true economic costs of providing the commodity has the advantage of restricting the demand of consumers and therefore of minimizing over-investment in water supply facilities.

When determining the least cost solution to supply the estimated demand for water the costs to be considered for each scheme include:

(a) initial costs of construction;
(b) recurrent operating costs; and
(c) costs of augmenting the water supply during periods of shortage.

It would perhaps be convenient to provide a set of guidelines which would simply indicate the situations where rainwater harvesting schemes are the least cost solutions to supply water. Unfortunately no such simplistic guidelines are possible. Every situation requires individual analysis.

It is possible, however, to indicate factors which result in relatively low total costs for precipitation harvesting schemes in any specific area. These include:

(a) high average precipitation (reduced catchment area);
(b) little seasonal or annual variability in precipitation (reduced storage volumes);
(c) available roof areas to serve as catchment areas at no cost to water supply schemes; and
(d) low land costs.

Situations where many of these factors exist are liable to be wet rural areas where all buildings have impermeable roofs (simply to waterproof them) which can be used to catch rainwater. But in such conditions the alternative forms of water supply such as rivers and lakes are also likely to be inexpensive and common, thus the development of surface and groundwater sources will also be relatively cheap.

Rainwater harvesting schemes will also be attractive where the costs of alternative water supplies are relatively high, for example in areas where no fresh surface water or groundwater exists. Such circumstances exist on many islands, such as Bermuda; although saline water conversion may be competitive for large-scale water demands in maritime areas.

Water harvesting from roof catchments has been practised for many years in various African countries, however, on a very small scale. The construction of houses with permanent and suitable roof catchments has an added advantage in that one draws good quality water ready for use. Thus no water maintenance costs are involved, and the rural people would not have to pay for any water collected from their own houses. It would be much easier and cheaper to organize people in the rural areas in self-help groups for purposes of building the appropriate structures and the storage tanks than organizing a conventional water supply system. Self-help methods would provide more than adequate water for people in the humid and subhumid parts of Africa,

Middle East, Asia and Central America.

It will also raise the standard of living of the people by providing good housing facilities, and minimizing surface runoff and other environmental hazards, such as erosion, sedimentation and flooding.

TECHNIQUES FOR EVALUATING GROUND CATCHMENT PERFORMANCE

Ree *et al.,* (1971) gave a hydrologic design method for determination of the size of storage tank and catchment for rainwater harvest schemes. Their method is based on an analysis of occurrence of minimum rainfall. They assumed a catchment of butyl rubber having high runoff efficiency. Illustration of the design of a water harvesting system taking untreated catchment would also have been useful. Frasier (1975) used a computer program to arrive at a better combination of storage and catchment size structures. He took into consideration all possible combinations of:

(a) rainfall patterns;
(b) water requirements; and
(c) unit construction cost.

Smith (1975) pointed out that Frasier (1975) should have included three more items in his model:

(1) probabilities of length of drought experienced;
(2) cost of supplying water from some other source (hauling); and
(3) storage performance for certain patterns of rainfall.

Hanson *et al.,* (1975) developed a computer procedure for simulating storage of water in ponds based on hydrologic information, contributing area, number of cattle, length of grazing season, seepage estimates and evaporation rates.

EVALUATION OF INTEGRATED SYSTEMS

Perhaps the greatest benefits in rainwater harvesting may come from a multi-use approach, i.e., the use of integrated systems. As the term suggests, integrated systems are those which combine two or more water harvesting techniques in a complementary manner. We, in fact, have already reviewed some systems which exhibit integrative characteristics. For example, the experiment in which a salt-treated catchment was combined with a water storage tank (figure 4-6) illustrates a system in which cropped waterways and stored excess runoff are integrated. Therefore it may be profitable to describe a number of experiments in which the combination of different water harvesting methods appears to have been particularly effective. For example, experimentation with water harvest in agriculture has suggested that, above a set minimum, the distribution of rainfall is more important than total rainfall. The conclusion that some researchers draw from this is that successful water harvesting which is dependable enough for effective commercial agriculture must be combined with efficient water storage in most semi-arid regions. In the example illustrated in figure 4-6 harvested runoff is stored and then pumped back to water the plants during dry spells. This technique has been called a 'water harvesting agri-system' (Cluff, 1978; Morin and Matlock, 1975).

Recent research on evaporation control has suggested that the reduction of surface area of a reservoir is an effective means of reducing evaporation. Further research determined that by dividing a conventional reservoir into compartments and by transferring water among these compartments, substantial evaporation control can be achieved by reducing the total surface area of the reservoir. Thus a three compartment system can reduce the surface area of a reservoir. The number of compartments, as well as their depth and size depend on the particular conditions of a given watershed. In addition to effective evaporation control, compartmented reservoirs reduce loss from seepage. This method can be applied to existing reservoirs as well as new ones. In fact, if the slope is greater than three to four per cent, a gravity-fed compartmented reservoir can be devised. However, a pump is usually required to transfer water from one compartment to another. Portable pumps are available commercially (the smaller 3.5 HP pumps costing approximately US $ 800). Since these are required only a few times a year, one pump can service numerous compartmented reservoirs (Cluff, 1978).

Compartmented reservors have been constructed in both the USA and Mexico. In the American southwest compartmented reservoirs have been integrated into water harvesting agri-systems on the Mavajo reservation in northern Arizona. In addition, water harvesting agri-systems including compartmented reservoirs have been used to rehabilitate abandoned farmlands by cultivation. The total cost (1978) of establishing and maintaining one plantation acre is US$ 1,608 (Cluff and Foster, 1978).

In Mexico, over eleven compartmented reservoirs have been constructed in the State of Coahuila. Some serve as livestock reservoirs; others are used for agricultural purposes. One of the most interesting experiments was conducted on the Ejido San Francisco del Barrial near Parras, Coahuila (Gavade *et al.*, 1976). This farm has a population of 350 inhabitants who collect candelilla wax, farm, and raise livestock on a total area of 2,400 hectares. Soils in the area are highly saline and, in places, waterlogged. Annual rainfall is between 200 and 250 mm. Rainfall falls in short hard storms during the summer months, but is quite unpredictable. Additionally, groundwater sources are available, but the water is saline. The installed water harvesting agri-system consisted of a 100 hectare water catchment, a 20 hectare orchard, collector drains, a three compartmented reservoir, and a supplementary water supply from a saline well. The 100 hectare catchment provided an additional water harvest collection area. The area was cleared and compacted and given a one to two per cent slope towards the reservoir. It is estimated that runoff from this area may be improved further (resulting in over 90 per cent runoff) by applying candelilla wax and natural fibre to the surface. The cost would be approximately five to ten pesos (US $ 0.20 to 0.40) per square metre.

The 20 hectare orchard serves as the main catchment. The surface has been cleared, shaped into collector terraces and waterways, and compacted. The nine metre wide collector terraces alternate with two metre wide waterways. The latter serve as cultivated micro watersheds. The waterways, which also serve to conduct runoff to the compartmented reservoir, were planted with drought-tolerant high-value crops including pistachios, grape vines, olives, almonds and also with subsistence crops such as beans and corn.

Preliminary observations indicate that the system is producing water of excellent quality. Furthermore, it is anticipated that operation and maintenance costs will be low (Gavade *et al.,* 1976). An estimate of the water budget for the entire system indicated that in an average rainfall year there should be enough water for consumptive use of the orchard as well as for supplementary irrigation of the dry farm area. In dry years, the water harvesting system may not provide sufficient runoff for either the orchard or the storage reservoir. In this case, well water may be used as a source of supplementary irrigation. So far salinization has not become a problem.

COMPARATIVE EVALUATION OF GROUND CATCHMENTS

Over the past thirty years a wide variety of experimental techniqes for the collection and storage of runoff water have been developed. As most of these techniques have been developed under experimental conditions in the USA, a certain measure of caution is required in the translation of the available benefit/cost calculations to other countries. At the outset, it can be stated that no 'best' or 'most effective' method exists, because local social and physical conditions are so varied. In general, the best methods are those that produce sufficient and dependable supplies of harvested water at the lowest cost. However, a method that is efficient and cost-effective in one setting, could be an inappropriate failure in the next. Keeping this in mind, we can review the effectiveness of the major categories of experimental water harvesting.

Vegetation Management

Although more research must be conducted on the strategy of vegetation management, initial results are not encouraging. Experimental research in Arizona indicates that grass cover is not as effective as other methods in inducing runoff. Moreover, experience in California suggests that the conversion of brush cover to grass may bring about severe soil slip erosion on steep slopes and flooding in low areas.

Land Alteration

Land alteration is probably the simplest and least costly of the experimental techniques. However, it is usually a flexible strategy which is easily integrated with other water harvesting techniques as various types of surface treatment, or water storage. These attributes are clearly demonstrated in the concept of harvesting water from highways, as well as the simple, inexpensive desert contour strips and conservation bench terraces. Both these forms of land alteration increase available water and crop yields significantly. However, one important drawback exists in the form of the unpredictability of the water harvests. Variable yields and occasional crop failure are an unavoidable part of this technique. This makes simple forms of land alteration unattractive for some crops (fruit tree crops) or for those farmers requiring a high degree of control and dependability.

Chemical Treatment and Surface Covers

The wide array of surface treatments and covers provide varying degrees of effective runoff, durability at varying costs. In general, chemical treatments and covers are much more effective at inducing runoff than vegetation management or land alteration; however, in many cases their cost and limited durability make them an unattractive option. The rising cost of petroleum and petroleum products will make this drawback even more significant in the future. This applies to use in water harvest catchments as well as to use for water storage facilities and evaporation control; although cost differentials can be more significant in the former case.

Integrated Systems

Integrated systems successfully combine the most effective features of a number of water harvesting techniques. This is particularly the case where land alteration techniques are combined with inexpensive soil treatment and provided with a backup compartmented reservoir storage system. In this fashion, the simplicity and low cost of the former technique are maintained while the drawbacks of undependable yields are minimized.

Constraints

All experimental water harvesting techniques exhibit important constraints. These constraints will become even more evident if the techniques are widely promoted for use in arid regions. The most significant constraint to consider in vegetation management is its possible environmental consequences. In fact, all forms of simple unassisted land alteration are subject to variable yields and crop loss during poor rainfall years. Soil erosion is a potential danger in land alteration schemes. With regard to chemical treatments and covers, cost is the most important constraint. Moreover, the component materials and equipment may not be available in many developing countries. Often the quality of water provided by some of these methods is fit only for agriculture and livestock, not for human consumption.

The constraints exhibited by different integrated systems depend on the constraints of their component techniques. Thus, an integrated system including desert contour strips may increase soil erosion, or the specification of an expensive pump may make an entire project unfeasible. However, as we have seen in the example of the water harvesting agri-system concept, a well-designed integrated system uses the advantages of one component to minimize the constraints of another.

EFFECTIVENESS OF BUNDS AND HAFFIRS

Water harvesting technology using bunds is a very widely applied technology in many regions in India. Unfortunately, however, local training in technical skills has not kept pace with the application of this technology, with the result that the maintenance and operation of bunded tanks has to be shouldered more and more by centralized agencies of the government. The loss of storage capacity in these tanks due to silting was revealed during a study sponsored by the Planning Commission of India (figure 10-7).

During the survey of 121 tanks in Tamil Nadu, 10 per cent were found to be silted up by more than half, and 50 per cent of the tanks had lost about 25 per cent of their storage capacity. If this is a typical representation of the situation it is reasonable to assume that it also represents an equilibrium situation, with old tanks at the left end of the graph nearing 100 per cent silting and new tanks at the right end, with the rest of the tanks gradually following a left-moving trend. Since tank sites have almost all been used up in peninsular

India very little increase will take place in the tank population. A time will therefore come when the right end of the graph will begin disappearing and the number of tanks in use will decrease progressively.

Reference has already been made to the fact that silt was being removed from the tank beds earlier by the users. This used to be done under a feudal system known as 'Zamindari', where large areas of land were owned by a 'Zamindar' (landlord), but cultivated by local villagers (tenants) on the basis of sharing the crop. When the tank bed dried in summer, the tenants were all required to remove the silt and spread it on the fields as part of their duties. This was more or less slave labour, and the Zamindari system was abolished soon after India gained independence. No alternative system, however, was introduced for desilting the tanks. With the advent of chemical fertilizers and the increasing cost of transporting the silt, the latter soon lost its attraction as manure. Loss of storage capacity due to silt accumulation has now become a serious problem. However, these tanks may still be suitable where other technologies are not. Bunded tanks can be built at low cost where the terrain is undulating, so that the bund is short. Both bunded tanks and haffirs (dug tanks) can be easily designed and constructed so that relatively large volumes of water can be stored submerging small areas of land.

Haffirs in many semi-arid regions, e.g., in Africa, have much to offer because of their possible use in integrated systems for crop and range management.

ASSESSMENT OF GROUNDWATER RECHARGE

Storage loss from silting, evaporative loss, seepage loss, danger of overtopping, and breaching, are all disadvantages to bunds and haffirs (dug tanks). In place of these systems it may be that harvested rain and stormwater in future years will simply be used to recharge groundwater. It would be advantageous to assess present methods used to store harvested water by recharge. For this purpose in India percolation tanks, submergence tanks and inundation canals have all been assessed.

Percolation tanks are especially suitable in undulating terrain where the soil is porous or rocky, so that the stored water percolates underground through soil pores or fissures in the rock to raise the water table. Full advantage can be derived from them if the percolation is rapid compared to evaporation. Since no long conveyance channels are necessary, there is very little loss of water by seepage. The water stored underground is not subject to evaporation either. A small loss occurs due to the flow of groundwater itself, whose velocity is of the order of 2 to 3 km per year in India.

Submergence tanks (ahars) are especially suitable in flat terrain where large areas can be submerged by a low bund, and where the soil retains enough moisture after the ahar is emptied to enable a crop to mature without further irrigation. Alluvial plains therefore offer the best sites for ahars. Advantages of ahars are that the loss of water by evaporation is small because of the low detention time, no cultivable land is lost under storage and there is no silt problem.

COMPARISON OF THE EFFECTIVENESS OF ANCIENT HARVESTING TECHNIQUES FOR AGRICULTURE

Several ancient water harvesting techniques have been reviewed in this book. Contour terraces and bordered gardens which are forms of rainwater harvesting, and silt traps and check dams which are stormwater harvesting techniques have been assessed. Their relative effectiveness most probably depended on the fact that, as practised by ancient cultivators, these techniques were essentially complementary, rather than competing strategies, within integrated water management systems.

Each technology was suited to maximize the moisture and runoff potential of particular topographic features, as in the case of contour terraces on slopes and silt traps in gullies. In this sense, each technology is unique and comparisons of their respective effectiveness is inappropriate.

Moreover, we often find these technologies acting as component parts of larger water harvesting complexes. Thus, irrigation water diverted from check dams enhances the yields of bordered gardens, and silt traps are found on sites both above and below check dams. In the former case they help protect the check dam from torrential flash floods; in the latter they benefit from excess water which is diverted to them by means of spillways and canals.

Keeping these facts in mind, we can turn to three aspects by which the comparative effectiveness of these techniques can be assessed.

First, with decreasing average annual rainfall there is a diminished capacity of simple rainwater harvesting strategies, such as contour terraces, to provide reliable crop harvests. Although many other variables enter the picture, it can be said that sites with less than 300 mm average annual rainfall can expect frequent crop failure on contour terraces unless these receive supplementary irrigation. On the other hand, stormwater harvesting techniques such as silt traps which concentrate runoff from a larger catchment rather than simply slope wash, have the capacity to yield crop harvests in years when contour terraces fail. However, a countervailing factor in this equation is the fact that the appropriate sites for silt traps are much more limited than slopes suitable for contour terracing.

Overall, the greater dependability of stormwater harvesting technologies with increasing aridity is reflected in the distribution of relic structures. In the southwest USA silt traps are much more important than contour terraces. The limitations on the expansion of an agricultural system dependent on this technology should be kept in mind. In the more humid latitudes further south these constraints are somewhat relaxed and we find a more expansive agricultural system capable of putting the resources of hillslopes to good use.

A second factor in the comparison between ancient rainwater harvesting technologies suggests that the productivity of bordered gardens was higher than that of contour terraces. This was due most probably to more intensive cultivation practised on the bordered gardens. Aside from the possibility that each garden served as microcatchment, there is no persuasive evidence that the bordering technique, itself, was a more effective water harvesting mechanism than contour terracing. It is more likely that the care lavished on bordered gardens reflected the necessity to conserve the limited irrigation water with which they were fed.

A third comparison can be made between two stormwater harvesting techniques, silt traps and reservoirs (bunds), which suggests that reservoirs were more versatile as they served both domestic and agricultural needs. A liability of reservoirs or bunds is the fact that they required a great deal of attention in order to prevent accumulations of alluvium that hampered their water storage capacity. Of course it is precisely this accumulation which is the objective of successful silt trap farming.

Several constraints to ancient water-harvesting practices can be inferred from the archaelogical record. They reflect the vulnerabilities of the different techniques. In the first place, we find the vulnerability to variations in rainfall, as described above. As we have seen, contour terraces are more vulnerable to both intra-seasonal variations and drought than other forms of water harvesting. Nevetherless, because all water-harvesting techniques are rainfall dependent, they imply a necessary degree of unreliability in subhumid environments. This constraint led ancient farmers either to favour sites with dependable water sources over those suited to water harvesting, or to complement water harvest cultivation with other sources of livelihood such as gathering or hunting.

The second important constraint concerns check dams more than it does contour terraces. Specifically ancient stormwater harvesting limited local farm production to the capacities of particular catchments. In several cases cultivators appear to have abandoned a scheme after having reached maximum development of numerous watersheds. Thus the limited scale of the system was unresponsive to the needs of a growing population.

Another constraint relates to water harvesting's labour requirements. As we have seeen, these were generally not significant at the construction stage. However, these systems appear to need continual maintenance and attention. Reservoirs must be cleaned out and reinforced against flash floods, silt traps must be built up to accommodate larger deposits of alluvium, and contour terrace ridges must be continually surveyed for damaging breaches. Water harvesting's sensitivity to maintenance is dramatically illustrated in certain cases where a significant decline in population led to the abandonment of hillside terracing and resulted in severe erosion which is still affecting some regions four centuries later.

CONTEMPORARY WATER HARVESTING FOR AGRICULTURE

Contemporary water harvesting technologies are associated overwhelmingly with small-scale subsistence farming, low incomes, and traditional crop complexes. A good example here can be taken from the characteristic peasant economies of Mexico and the Indian agriculturalists of the American southwest.

In the American southwest, we find residual water harvesting, the Indian populations have emerged from their encounter with nineteenth and twentieth century Anglo-American hegemony with much reduced homelands, a precarious control over the mineral and water resources within their reservations, a subordinate social status, a weakened cultural identity, and an impoverished economy. The myriad pressures of underdevelopment and marginalization have resulted in a decline or abandonment of the old livelihood patterns including subsistence agriculture, and an assimilation of many individuals to urban American society. For those remaining on the reservations the shift from farming and gathering has been mainly in the direction of wage labour, craft sales, and livestock raising. These groups now face new pressures as the US economy is making increasing demands on the energy resources located within their reservations.

It is difficult to make any broad generalizations regarding the capacity of water harvest agriculture to sustain peasant households in Mexico, as this varies greatly under different physical and social conditions. For example, one study estimating maize yields on terraces in a community in Tlaxcala calculated that almost 40 per cent of the annual maize harvest was available for sale after household needs were met (Patrick, 1977). On the other hand, another study documented the fact that 88 per cent of households in one community in the Mezquital Valley of Hidalgo could not achieve maize self-sufficiency on the basis of water-harvest farming involving both contour terraces and silt-trap check dams (Johnson, 1977).

Although most of this disparity may be accounted for by the difference in annual rainfall (Tlaxcala experiences approximately 300 mm more rain than Mezquital Valley), it is also important to note that in Tlaxcala the community population densities are approximately twice those of the Mezquital Valley communities.

Overall the occupational picture of the average Mexican water harvest farmer is a precarious one. This farmer as in the American southwest must frequently resort to alternative sources of income including livestock raising, crafts manufacture, and increasingly, wage labour. It seems probable that the spatial extent of water harvest agriculture has shrunk since historic times, substantially in the case of the American southwest, moderately in the case of Mexico, but it has expanded in Africa and China (chapter 11).

COMPARATIVE EFFECTIVENESS OF WATER HARVEST FOR AGRICULTURE

As with ancient water harvesting, in actual practice, the modern methods are complementary rather than competitive and are often found in combination with integrated moisture management systems. Also, as with ancient water harvesting, contemporary stormwater harvesting is more effective than rainwater harvesting in regions of higher aridity. Also, specific contemporary water harvesting systems vary considerably in terms of quantity and adequacy of harvesting potential since they are so highly dependent on site and situation.

Studies made of contemporary systems confirm most, if not all, of the observations, deductions, and assessments made in regard to their ancient counterparts. A consideration of sand dune and floodwater farming serves to highlight one important feature shared by contour terraces and silt traps, namely, their capacity to enhance soil and water conservation. Sand dune and floodwater farming techniques are either ineffective or, in the case of sand dune farming, even prejudicial in this regard. Contour terraces and silt traps are very effective measures to combat erosion and desertification.

Table 13-1 summarizes the comparative effectiveness of the five techniques reviewed in this section in terms of relative cost effectiveness under extremely arid conditions, reliability, and potential as a conservation measure.

The constraints exhibited by contemporary water harvesting techniques are numerous and, in some cases, significant. Contour terraces, silt traps, and reservoirs require a steady investment of attention and labour in order to provide for their maintenance and (in the case of contour terraces and silt traps) expansion. A decline in requisite labour inputs can lead to impaired yields and even to serious environmental deterioration. Construction of sand dune fields, silt traps, reservoirs, and floodwater fields is also contingent upon the presence of appropriate topographic and hydrographic features. These may exist in a very narrow range of sites, thereby limiting the diffusion of a given water harvesting technique or its expansion in a particular location. With the exception of contour terraces, most contemporary water harvesting systems are small-scale, usually serving the needs of individual households or

Table 13-1 The comparative effectiveness of harvest techniques

Harvest technique	Cost	Effective 200 mm	Reliability	Conservation Measure
Contour terraces	low	no	low	yes
Sand dune fields	low	partial/yes	low to moderate	no
Silt traps	moderate	yes	moderate to high	yes
Reservoirs	low to high	yes	moderate to high	--
Floodwater fields	low	yes	low to moderate	no

villages. Often, these systems already are capturing maximum feasible amounts of runoff.

Contour terraces, sand dune fields, and floodwater fields cannot be relied upon to provide for crops in drought years. Hence, they must be supplemented, or play a secondary role within a production system that incorporates more reliable forms of livelihood. Silt traps and reservoirs exhibit this constraint to a lesser degree. While most water harvesting is actually beneficial in terms of soil and water conservation, two techniques—sand dune and floodwater farming—may result in soil erosion, raising questions about the long-term stability of production based on these practices.

Even with constraints in the form of silting difficulties and unreliable harvests, most contemporary water harvesting techniques have been able to make effective use of the labour and locally available resources in peasant communities. Additionally, the Mexican government has found that labour-intensive conservation projects featuring contour terraces and silt traps are relatively inexpensive and effective.

CULTURAL IMPLICATIONS

In the American southwest the definite trend away from water harvesting has constricted its use to a fraction of its previous territory. Nevertheless, over the past thirty years, water harvesting has been used by the Hopi, Zuni, Navajo, Papago, Mojave, Yuma, Cocopa, and Maricopa groups inhabiting the states of Arizona and New Mexico. In addition water harvesting techniques have endured for centuries among some groups, and moreover, are diffused widely throughout the territory and ethnic populations in China, India and Mexico. This is a testimonial to the usefulness and versatility of this technology. However, it is important to note that groups practising large-scale capital intensive agriculture have by-and-large shunned traditional water-harvesting methods and have opted for other technologies.

Most water-harvest systems can be constructed and extended in an incremental fashion so that investments in labour and resources can be spread over long time periods. However, the extension of water harvest agriculture to cover entire slopes also requires a certain level of cooperation and integrated management. For example, poorly maintained contour terraces, silt traps and check dams deteriorate and erode and thus pose a hazard to downslope structures. Therefore, a series of water harvest structures which are owned by different households imply a community level of cooperation in order to maintain the system as a whole. In many regions of the world where rain and stormwater harvesting is practised, future efficiency will probably demand an integration of different systems.

CHAPTER 14

Recommendations and Criteria for Rainwater Harvest Pilot Projects

OVERALL, it would appear that (with a few exceptions) the experimental water-harvesting techniques that have been developed over the past few decades, have not been widely diffused even in the USA, their country of origin. It is not clear why this is the case, as many have been demonstrated by researchers to be efficient and cost effective. In most cases, where experimental water harvest systems have been adopted or promoted, it has been by government agencies (both US and Mexican) and not by private users or producers.

The following recommendations are made on the assumption that, for the near future, this situation will continue, and that international and national agencies will take the lead in promoting the most promising of the experimental technologies.

In general, those water harvesting technologies which maximize local resources, materials and labour are considered to be superior to technologies which rely on imported materials, equipment and high technology.

Domestic Water Supply

ROOF CATCHMENTS

1 More information is needed on roof catchment techniques used on islands especially in the Caribbean. These are of interest because the methods have been in use for many years. A good example is Bermuda where the law requires rooftop water collection, yet very little published information is available.

2 Roofing materials and construction methods in developing countries should be selected with the idea of rain harvesting in mind.

3 Research is needed into appropriate materials for roofing and ground catchments in developing countries.

STORAGE OF HARVESTED RAINWATER

1 Water storage has come a long way, but research and field testing of novel combinations of storage and water treatment should be tried. More information especially is needed on water quality during long-term storage in concrete tanks (as used in the Caribbean) or underground storage wells (as used in China).

2 More experimentation is needed on storage containers at the intermediate technology level. Innovations such as burial of large storage vessels to increase their life span could be adopted.

3 The use of indigenous soils and plants in the purification of rainwater through flocculation and settling should be tested and developed for application under appropriate conditions.

4 Manually operated, simple low-cost pumping devices should be encouraged for widespread application in drawing water from underground rainwater storage containers which serve one or a few family households.

5 Storage and distribution of harvested rainwater should observe the principles outlined in the *Action Plan Against Schistosomiasis,* namely:
   (a) prevention of physical contact with stored water (through 'physical' or 'distance' barriers);
   (b) catchment area protection;
   (c) additional in-line storage for 1 to 2 days supply for the community;
   (d) separate laundry and bathing facilities.

6 Where no water is to be wasted, two containers must be used, so that the first flush of polluted water goes into one container which is used for washing. The clean water is collected in the other for drinking and cooking.

7 Aquatic insect development, especially mosquitoes, is prevented by use of a kerosene film.

DEW AND MIST HARVESTING

1 Dew and mist harvesting should be investigated as a source of additional water supply. In this respect more information is needed on actual dew harvesting applications in use in countries like Guinea-Bissau.

Agricultural Water Supply

GROUND CATCHMENTS

1 Among the most promising experimental techniques here are the integrated systems of water harvesting agri-systems which combine simple (perhaps sodium treated) shaped, compacted earth catchments with gravity-fed compartmented reservoirs.

2 Desert contour strips, contour bench terraces and other forms of shaped compacted earth catchments, are also to be encouraged.

3 Sodium-treated, compacted earth catchments are of use especially for livestock watering.

4 Gravity-fed, compartmented reservoirs are the most desirable for ground catchment storage, while stone or sand-filled reservoirs are also useful in terms of conserving stored water.

5 Compacted earth reservoirs perhaps treated with sodium bentonite are useful in areas suffering from seepage losses.

6 On islands with very porous substrata, such as coral, the Manda Island Case Study (chapter 6) can be used as a model.

HAFFIRS

1 Haffirs should be encouraged providing the following conditions are present:

(a) minimum mean of 80 mm per rainy season, if this coincides with the cold period of the year, more than 80 mm if it occurs during the summer when evaporation is greater;

(b) catchment area containing impermeable, or crust-forming, or poorly draining soil (as in southern Sudan);

(c) cultivated area soils with high water-storing capacity, 1.5 to 2 m deep; and

(d) salinity in the cultivated soil not exceeding 2 to 3 per cent.

2 Evaporation problems should be minimized by digging deeper excavations and by using compartmented reservoirs to minimize exposed surface area.

3 Sand-filled wells or other intermediate technology should be used to provide a source of clean water for domestic use in the deep end of haffirs.

BUNDED TANKS

1 In cases where land is still available to build bunded tanks, simple compacted earthen dams and compacted earth watersheds should be constructed where topography, soil conditions and space allow.
2 In regions such as India where new bunded tanks are not likely to increase in number, the existing tanks should be made more efficient by:
   (a) optimization of storage capacities in relation to evaporation, depth of storage and agricultural operations;
   (b) systems analysis of tank management, particularly where tanks are built in series;
   (c) measures to reduce evaporation and seepage from the distribution system at low cost;
   (d) evaluation of various techniques of desilting;
   (e) measures to reduce silting, such as construction of upstream check dams to trap silt, and vegetation development in the catchment to minimize erosion;
   (f) operation of the tanks like submergence tanks (ahars) where conditions permit;
   (g) plantation of trees and natural fence plants around those tanks to be used as domestic water supply;
   (h) as with haffirs, the use of sand-filled wells should be encouraged to provide a source of clean domestic water in multi-use tanks.

MICROCATCHMENTS

1 The use of microcatchments in semi-arid regions should be encouraged with the following factors in mind:
   (a) the water-use characteristics of the plants to be grown;
   (b) their potential yields;

(c) their ability to resist drought;
(d) whether the soils in the cultivated area can store enough water to mature the crop; and
(e) the amount of evaporation from the soil surface.

2 If these conditions are not present, then the local terrain must be modified, e.g., by increased tilling and additions to the soil to increase its water-storing capacity, or by sealing the çatchment soil to increase its impermeability, etc. All of these changes will increase local costs.

FLOODWATER FARMING

1 The principles of water spreading practised by some floodwater farmers are attractive and deserve consideration. However, as traditional water harvest systems are in full retreat in this area, the efficiency of this technique is difficult to disentangle from the socio-economic context of its use. Therefore because of its unreliability and environmental vulnerability this technique is not recommended.

SUBMERGENCE TANKS ('AHARS')

1 In place of the unreliable floodwater farming technique it is recommended that a more controlled method of systematic flooding be used. This method is referred to in India as a submergence tank (or 'ahar') and should be made more efficient by research into:

(a) stabilization of slopes of the ahar bund at low cost;
(b) systems analysis of ahar operation for flooding, storage and/or irrigation and in series (storage alternating with flooding);

(c) such systems analysis should use mathematical models which consider:
  (i) optimum time of detention of water in the ahar,
  (ii) rates of percolation and evaporation,
  (iii) capillary moisture capacity,
  (iv) depth of water table,

(v) constraints imposed by time-bound agricultural operations such as sowing, etc.,
(vi) volume of water to be stored in the root zone of the soil, and
(vii) other factors such as salt content and plant nutrients present in the ahar soil after flooding.

PERCOLATION TANKS

The storage of harvested water by groundwater recharge holds much promise and should be encouraged especially in India where the bunded tank system will gradually become obsolete. In order to determine the efficiency of groundwater storage it is recommended that:

1 A comparative study of the rates of percolation and evaporation must be made. This, together with other parameters like storage capacity below the water table, withdrawal rates, soil permeability and so on, could help formulate a model which determines the optimum capacity of the percolation tank, subject of course to local constraints.
2 Verification of the model through the determination of increase in the actual yields, rise in water table, and distance to which the influence of the percolation tank extends, etc.

DEW HARVESTING

1 Research and information is needed on the use and potential of dew and mist in agriculture.
2 Special attention should be paid to the use of gravel fields, meshes, poles and even vegetation, as interceptors of dew and mist.

LAND CONTOUR SYSTEMS (TERRACING, SILT TRAPS, CHECK DAMS)

1 Contour terraces increase the moisture-carrying capacity of gentle slopes and thereby enhance the productivity of slope contour fields.

Available studies indicate that this technique achieves these results at remarkably low costs. Moreover, the technology is simple, the materials local, and the approach lends itself to labour-intensive public projects favoured by many developing countries. Contour terraces are particularly suited to subhumid regions receiving rainfall averaging over 400 to 500 mm a year.

2 Although sites appropriate for silt traps are less numerous than those for contour terraces, most available evidence suggests that this technology is a highly effective means of concentrating water and soil resources for agriculture and conservation. Moreover, silt traps have proved to be more effective and reliable than other harvesting systems in highly arid regions. However, scale limitations may preclude their use by large-scale farmers. Silt traps, even more than contour terraces, are explicitly suited to labour-intensive, grass roots community development projects.

3 Check dams are in many instances suitable for wider application. Serious problems arise in terms of their suitability for domestic uses, yet moderate-size check dams located in favourable sites provided with adequate maintenance should ensure valuable supplies of harvested stormwater which can be used for agriculture and livestock.

INTEGRATED SYSTEMS

As the present day Chinese and relict Central American examples illustrate, successful harvesting, storage and utilization of rain and stormwater are inseparable from the integrated management of small watersheds. In all small watersheds where water conservation is needed, comprehensive measures must be taken including:

1 Transforming some slopes into terraced or semi-terraced fields.
2 Planting trees, shrubs and grasses on uncultivated steep slopes.
3 Building series of silt-traps or check dams in gully channels to retain silt and water and create new farmland.
4 Diversion of stormwater for use on farmland ('warping').

If these measures are undertaken in an integrated fashion, soil and water resources can be conserved and agriculture yields can be significantly augmented.

In general terms land alteration methods are the simplest and least costly of all the alternatives. Moreover, they maximize the use of local manpower and materials, and use a technology that is flexible and can easily be integrated with various types of catchment treatments, such as soil treatment, plastic membranes, etc., if so desired.

PILOT DEMONSTRATION SITES

The United Nations Environment Programme has an interest in providing for a series of Pilot Demonstration Sites where all of the technologies recommended above can be demonstrated at several sites in different regions of the world. In selecting the sites for the location of future water harvesting demonstration projects the following six criteria should be kept in mind:

Climatic Zone

(a) Priority should be given to sites located in subhumid regions receiving up to 500 mm rainfall per year.
(b) The site should exhibit a marked seasonal concentration of rainfall, i.e. 74 per cent rainfall falling within a six month period.

Demography

(a) Priority should be given to small and medium size rural settlements within which a range of technologies applicable at the individual household, the community and the multi-village level can be demonstrated.
(b) High densities and competing uses may preclude successful adoption of community-level water harvesting techniques; however, this should not preclude pilot demonstrations of techniques suitable at the individual household level (e.g., roof-top or courtyard water collection and small-scale storage devices).

Level of Socio-economic Development

(a) Priority should be given to developing countries.
(b) Priority should be given to needy populations.

Urgent Need

(a) Priority should be given to areas exhibiting an urgent need for the provision of rural water supplies—either domestic or agricultural—where shortages have been clearly identified.
(b) Priority should also be given to areas where potential shortages can be predicted.

Cultural

(a) A high degree of sensitivity to local cultural norms and values should be maintained.
(b) The site should be located within a zone of cultural homogeneity in order to maximize the diffusion potential of the new practice.
(c) Pilot demonstration projects should be distributed equally over the world's regions.

Evaluation

The evaluation at each site of all technologies mentioned in this book would be a difficult job. The problem is basically that a given technology might be acceptable to an African community at one site but may not be acceptable to a Central American, European or Asian community at other sites. Consequently, the final evaluation of individual technologies will depend on the region, the people, local materials, etc. As a guide we have drawn up two tables for use at the Pilot Demonstration Sites. This will enable individual projects to score technologies on a local basis using some relative score, say 1 to 10. Table 14-1 is designed for evaluating rainwater harvesting at project sites where the water will be used for domestic purposes. Table 14-2 is designed for those sites where the water will be used for agriculture.

Table 14-1 Evaluation criteria for rainwater harvesting at a project site where the water will be used for domestic purposes

TECHNOLOGY	EVALUATION CRITERIA									
	Rainfall quantity	Siting and scale	Reliability	Cultural accept-ance	Cost (initial)	Maintenance required and durability	Cost effectiveness	Maximizes local input	Health aspects	Overall Score
Interception										
1 Roof										
2 Courtyard										
3 Ground										
Natural										
Prepared										
Selection										
1 Channel or Funnel										
2 Set. Tank										
Storage										
1 Jars										
2 Storage wells (above)										
3 Cistern										
4 Underground tank										
Quality										
1 Chemical										
2 Nat. pro.										
3 Covers										
4 Filtration										
Retrieval										
1 Bucket										
2 Hand pumps										
3 Endless chain										

Table 14-2 Evaluation criteria for rain and stormwater harvesting at a project site for agricultural use

TECHNOLOGY	EVALUATION CRITERIA												
	Effective rainfall	Siting and scale	Reliable harvests	Cultural accepta-bility	Cost (initial)	Mainte-nance	Cost effect-iveness	Maximizes local inputs: 1. labour 2. materials	Conserva-tion measures	Associa-ted erosion problems	Associa-ted health problems	Water quality	Overall score
1 *Contour Farming*													
Type 1 (Terracing)													
Type 2 (deep furrows)													
Type 3 (other)													
2 *Silt Traps*													
Type 1 (simple)													
Type 2 (warping)													
Type 3 (other)													
3 *Check Dams*													
Type 1 (natural)													
Type 2 (prep.)													
Type 3 (haffirs)													
4 *Floodwater Farming*													
Type 1													
Type 2													
Type 3 (other)													
5 *Micro-catchments*													
6 *Dew harvesting*													
7 *Other*													

BIBLIOGRAPHY

Anaya, M.G., and Tovar, J.S. (1975), Different soil treatments for harvesting water for radish production in the Mexico Valley, in: *Proceedings of the Water Harvesting Symposium,* US Department of Agriculture, Agricultural Research Service.

Anon. (1935), *Aerial wells and soil moisture.* Tropical Agriculture, 13(2), 34.

Anon. (1961), *Vegetation studies—measurement of atmospheric moisture.* Record of Research, East African Agricultural and Forestry Research Organisation, Kenya.

Anon. (1963), Dew: facts and fallacies. *The Water Relations of Plants,* ed. by Rutter and Whitehead, Blackwell, Oxford.

Anon. (1966), *Committee on Plan Projects,* Irrigation Team, All India Review of minor irrigation works based on Statewise Field Studies, planning commission, New Delhi, 1966.

Anon. (1967), *Water in dry places.* Engineering, 204 No. 5297.

Anon. (1978), Encyclopaedia Britannica.

Anon.(1978), *Sedimentation studies in streams and reservoirs.* Annual Review 1978, Research Schemes Applied to River Valley Projects, Central Board of Irrigation and Power, New Delhi, India.

Arvidson, I. and Hellstrom, B. (1955), *A note on dew in Egypt:* Bull. Institute of Hydraulics, 48, Stockholm.

Beattie, B. (1969), Managing Forest Land for Water Protection, in: *Proceedings of University Seminar on Pollution and Water Resources.*

Boyer, D.G. and Cluff, C.B. (1972), An evaluation of current practices in seepage control, in: *Hydrology and Water Resources in Arizona and the Southwest,* Vol. 2, Proceedings, Meetings of the Arizona Section, American Water Resources Association and the Hydrology Section, Arizona Academy of Science.

Bradfield, M. (1971), *The changing pattern of Hopi agriculture.* London: Royal Anthropological Institute of Great Britain and Ireland.

Bryan, K. (1929), *Floodwater farming*: Geographical Review, 19; 444-456.

Castetter, E. and Bell, W. (1942), *Pima and Papago Indian agriculture.* Inter-Americana studies 1. Albuquerque: University of New Mexico Press.

Charlton, T. (1970), Contemporary agriculture of the valley: W. Sanders *et al., The Teotihuacan Valley Project Final Report,* Vol. I, The Natural Environment, Contemporary Occupation and 16th Century Population of the Valley. Pennysylvania State University Occasional Papers in Anthropology.

Chow, V.T. (1964), *Handbook of Applied Hydrology.* Macmillan, N.Y.

Cluff, B. *et al.* (1969), *Development and evaluation of water harvesting systems,* Project Competition Report, OWRR Project A-011-Ariz., Tucson: University of Arizona.

Cluff, B. *et al.* (1972), *Development of economic water harvest systems for increasing water supply phase II.* Project Completion Report. OWRR Project B-015-Ariz. Tucson: University of Arizona.

Cluff, B. (1975), Engineering aspects of water harvesting research at the University of Arizona, in: *Proceedings of the Water Harvesting Symposium,* U.S. Dept. of Agriculture, Agricultural Research Service.

Cluff, B. (1978), *The use of the compartmented reservoir in water harvesting agri-systems,* Paper presented at the International Arid Land Conference on Plant Resources, Lubbock, Texas.

Cluff. B. and Foster K. (1978), *Water harvesting agrisystem to grow jojoba on developed idle farmlands,* Paper submitted to the *Arizona Farmer Rancher.*

Cooley, K. (1975), *Evaporation suppression for conserving water supplies,* Proc. of the Water Harvesting Symposium, U.S. Dept. of Agriculture, Agricultural Research Services.

Cooley, K. *et al.* (1975), *Water harvesting: state of the art,* Watershed Management Symposium of the ASCE Irrigation and Drainage Division.

Cooley, K. *et al.* (1976), Stock-water harvesting with wax on the Arizona strip, in volume 6, *Hydrology and Water Resources in Arizona and the Southwest,* Prooceedings, Meetings of the Arizona Section, American Water Resources Assn. and the Hydrology Section, Arizona Academy of Science.

Cushing, F. (1920), *Zuni Breadstuff. Indian Notes and Monographs,* Vol. VIII. New York: Museum of the American Indian.

Deacon, E.L., Priestly, C.L.H.B., and Swinbank, W.C., (1958), *Evaporation and the Water Balance.* Climatology. Reviews of Research, Arid Zone Research, No. 10—UNESCO, Paris.

Dedrick, A. (1973), *Evaluating outdoor weatherability of butyl rubber sheeting under stress,* Transactions, ASAE 16:769-772.

Dedrick, A. and Williamson, W. (1973), *Operation, serviceability and material evaluation of raintraps on the Fishlake National Forest,* 1960-1971. Research Report 4, Ag. Exp. Sta., Utah State University.

Dedrick, A. (1975), *Storage systems for harvested water,* Proceedings of the Water Harvesting Symposium, U.S. Dept. of Agriculture, Agricultural Research Service.

Dedrick, A. (1976), *Water harvesting–modern application of an ancient method.* Civil Engineering, ASCE, October.

Deshmukh, M.T. (1979), *Critical Appraisal of work done on water harvesting systems,* Proceedings of the 3rd World Congress on Water Resources, Vol. I, pp. 256-265.

Dutt, G. and McCreary, T. (1975), *Multipurpose salt treated water harvesting system,* Proceedings of the Water Harvesting Symposium, U.S. Dept. of Agriculture, Agricultural Research Service.

Duvdevani, S. (1957), *Dew Research for Arid Agriculture,* Discovery 18:330-334.

Evans, C. *et al.* (1975), Opportunity for harvesting water from and along highways in rangeland areas of Wyoming, Proceedings of the Water Harvesting Symposium, U.S. Dept. of Agriculture, Agricultural Research Service.

Evenari, M., Shanan, L. and Tadmore, N. (1971), *The Negev.* Harvard University Press, Cambridge, Mass.

Fairbourn, M. (1975), *Field evaluation of microwatershed and vertical mulch systems,* Proceedings of the Water Harvest Symposium, U.S. Dept. of Agriculture, Agricultural Research Service.

Ffolliott, P. and Thorud, D. (1975), *Vegetation management for increased water yield in Arizona.* Technical Bulletin 215, Agricultural Experiment Station. Tucson: University of Arizona.

Fink, D. and Cooley K. (1973), *Water harvesting for improved grazing efficiency;* Proceedings of the Water-Animal relations symposium, Twin Falls, Idaho.

Fink, D. *et al.* (1973), *Wax-treated soils for harvesting water.* J. Range Management, 26: 396-398.

Fink, D. and Frasier, G. (1975), Water harvesting from watersheds treated for water repellancy; Reprinted from *Soil Conditioners,* copyright Soil Science Society of America, Inc., Madison, Wisconsin, 173-182.

Fink, D. and Mitchell, S. (1975), Freeze-thaw effects on soils treated for water repellancy, in Vol. 5, *Hydrology and Water Resources in Arizona and the Southwest,* Proceedings, Meetings of the Arizona Section, American Water Resources Assn. and the Hydrology Section, Arizona Academy of Science.

Frasier, G. (1975), *Water harvesting: a source of livestock water.* J. Range Management, 28 : 429-434.

Frasier, G. and Myers L. (1972), *Bonding films to soil surface for water harvesting.* Transactions of the ASAE. 15:909-911.

Frasier, G. and Schreiber H. (1978), *Rangeland forage rehabilitation by water harvesting,* Proceedings of the First International Rangeland Congress, Denver, Colorado.

Froebel, R. *et al.* (1978), *Asphalt-crumb rubber waterproofing membrane.* J. of the Irrigation and Drainage Division, Proceedings of the ASCE, No. IRI. Proc. Paper 13614 104:43-58

Gardner, H. (1975), *An analysis of the efficiency of micro-watershed systems,* Proceedings of the ASCE, No. IRI. Proc. Paper 13614 104: 43-58.

Gardner, H. (1975), *An analysis of the efficiency of micro-watershed systems,* Proceedings of the Water Harvesting Symposium, U.S. Dept. of Agriculture, Agricultural Research Service.

Gavade, S. *et al.* (1976), *System for managing saline and runoff water for fruit and crop production in arid regions of Mexico.* FAO-MEX 74/003 Technical Report.

Gindell, I. (1973) *A new ecophysiological approach to forest-water relationships in arid climates.* b.v. Junk Pub., the Hague.

Gleason, M.N., Gosselin, R.E. and Hodgen, H.C. (1957), *Clinical Toxology of Commercial Products.*

Griffin, E. and Dennis, H. (1969), *A Mexican corporate campaign in conservation.* The Professional Geographer, 21: 358-359.

Grover, B. (1969), *Harvesting Precipitation for community water supplies.*

Gruner, H.E. (1944), *Bewässerung - Anlagen - Boobachtungen und Erfahrungen beim Bau in Südlichen Landern,* Gebr. Leemann. Zurich.

Hack, J. (1942), *The Changing Physical Environment of the Hopi Indians of Arizona.* Papers of the Peabody Museum of Ethnology and Archaeology, Vol. 35, No. 1. Cambridge, Mass.: Peabody Museum.

Halacy, D.S. Jnr. (1966), *The Water Crisis.* E.P. Dutton and Co. Inc., N.Y.

Herold, L. (1970), *Trincheras and Physical Environmental along the Rio Gavilan,* Chihuahua, Mexico. Publications in Geography Technical paper No. 65-1. Denver: Dept. of Geography, University of Denver.

Hillel, D. (1967), *Runoff Inducement in Arid Lands.*

Hori, T. (1953), *Studies on fog in relation to fog preventing forest.* Tanne Trading Co. Ltd., Sapporo, Hokkaido, Japan.

Hubbard, A.J. and Hubbard, G. (1905), *Neolithic dew ponds and cattleways,* Longmans Green & Co., London.

Hursh, C.R. and Pereira,H.C. (1970), *Field moisture balance in the Shimba Hills, Kenya.* East African Agricultural Journal, 18 (4).

Jian, L. (1979), *Rain and stormwater harvesting in China.* Special Report to UNEP, Nairobi.

Johnson, K. (1977), *Do as the land bids. A study of Otomi resource-use on the eve of Irrigation.* Clark University: unpublished Ph. D. Dissertation.

Johnson, K. (1979), *Rain and stormwater harvesting in the USA and Latin America.* Special Report to UNEP, Nairobi.

Jones, O. and Hauser V. (1975), *Runoff utilization for grain production.* Proceedings of the Water Harvesting Symposium. U.S. Dept. of Agriculture, Agricultural Research Service.

Jones, P.E. (1979), Non-conventional water sources—a review of ideas and possible applications in Kenya. In: *The Role of Water in Development,* Kenya Academy of Sciences, Nairobi, pp. 49-56.

Jumikis, A. R. (1965), *Aerial wells–Secondary sources of water.* Soil Science 100 (2) 83-95.

Kashima, T. (1950), *Decrease of sea-fog density by a model shelterbelt.* IUFRO 12th Congress, Oxford. Papers, Volume-Band I, Section 11/9, pp. 48-49.

Kenyon, A.S. (1929), *The Iron Clad or Artificial Catchment.* Journal of the Dept. of Agriculture of Victoria.

Kerfoot, O. (1968), *Mist precipitation on vegetation.* Forestry Abstracts, 29 9-20.

Kirkby, A. (1973), The use of land and water resources in the past and present Valley of Oaxaca, Mexico. Vol. 1 in K. Flannery, ed., *Prehistory and Human Ecology of the Valley of Oaxaca.* Ann Arbor: Memoirs of the Museum of Anthropology, University of Michigan.

Kovacs, G. (1979), *Rain and stormwater harvesting in Europe.* Special report to UNEP, Nairobi.

Krosigh, S.V. (1971), *Die Bedeutung der Wasserwirtschaft für Afghanistan,* Die Wasserwirtschaft, Vol. 61, No. 6.

Lauritzen, C.W. (1967), *Butyl–for the collection, storage and conveyance of water.*

Martin, D.E. (1910), *Report on impervious water conservation catchments,* Eyre Peninsula, South Australia.

McBride, M. and Shiflet L. (1975), *Water harvesting catchments in the Safford District, Southeastern Arizona.* Proceedings of the Water Harvesting Symposium, U.S. Dept. of Agriculture, Agricultural Research Service.

Medina, J. *Harvesting surface runoff and ephemeral streamflow in arid zones.* Conservation in Arid and semi-arid zones. FAO, Rome.

Meigs, P. (1953), World distribution of arid and semi-arid homoclimates, in *Reviews of Research on arid zone hydrology.*

Meyers, E.L. (1969), *Precipitation runoff inducements.* Water supplies for arid regions.

Milne, L. and Milne J. (1964), *Water and Life.* Athenium Press, Boston.

Milthorpe, F.L. (1960), *The income and loss of water in arid and semi-arid zones. Plant water relationships in arid and semi-arid conditions.* Reviews of Research, arid zone research Number 15. UNESCO. Paris.

Monteith, J.L. (1957), *Dew,* Quart. J. Royal Meteorological Society. 83:322-341.

Morin, G. *et al.* (1973), *Desert strip farming—a way to make the desert green.* Progressive Agriculture in Arizona. 25:4-8.

Morin, G. and Matlock W., (1975), *Desert strip farming—computer simulation of an ancient water harvesting technique.* Proceedings of the Water Harvesting Symposium, U.S. Dept. of Agriculture, Agricultural Research Service.

Myers, L. (1967), *New water supplies from precipitation harvesting.* Paper presented at the International Conference on Water for Peace, Washington D.C.

Myers, L. *et al.* (1967), *Sprayed asphalt pavements for water harvesting.* Journal of the Irrigation and Drainage Division, ASCE, No. IR3, Proc. Paper 5413 93:79-97.

Myers, L. and Frasier, G. (1969), *Creating hydrophobic soil for water harvesting.* Journal of the Irrigation and Drainage Division, ASCE, No. IRI, Proc. Paper 6436, 95, pp. 43-54.

Myers, L. and Frasier, G. (1974), *Asphalt fibreglass for precipitation catchments.* J. Range Management, 27, pp. 12-15.

Nagel, J.F. (1956), *Fog precipitation on Table Mountain.* Quart. J. Royal Meteor. Soc. 82, pp. 452-460.

Ongweny, G. (1979), *Rain and stormwater harvesting in Africa.* Special Report to UNEP, Nairobi.

Orni, E. and Efrat, E. (1964), *Geography of Israel,* Israel Programme for Scientific Translations, Jerusalem.

Palmer, M.C. (1962), *Algae in Water supplies.* Ohio State Publications.

Patrick, L. (1977), *A Cultural Geography of the use of seasonally dry sloping terrain: The Metlepantli crop terraces of Central Mexico.* University of Pittsburgh: Unpublished Ph.D. Dissertation.

Prasad, R. (1979), *Rain and stormwater harvesting in India and the Far East.* Special Report to UNEP, Nairobi.

Rice, R. and Foggin, G. (1971), *Effect of high intensity storms on soil slippage on mountainous watersheds in Southern California.* Water Resources Research 7: 1485-1496.

Rohn, A. (1963), *Prehistoric soil and water conservation on Chapin Mesa, Southwestern Colorado.* American Antiquity 28 : 441-455.

Samuelson, P.A. (1967), *Economics: An Introductory Analysis.*

Sanders, W. (1965), *The Cultural Ecology of the Teotihuacan Valley.* Dept. of Sociology and Anthropology, Pennsylvania State University, Unpublished Report.

Sanders, W. *et al.* (1970), *The Teotihuacan Valley Project Final Report,* Vol. I. The Natural Environment, Contemporary Occupation, and 16th Century Population of the Valley. Pennsylvania State University, Occasional Papers in Anthropology.

Schreiber, H. and Frasier, G. (1978), *Increasing rangeland forage productivity by water harvesting.* J. Range Management, 31: 37-40.

Schwerdtfeger, P. (1979), *Rain and stormwater harvesting in Australia and the South Pacific.* Special Report to UNEP, Nairobi.

Sheppard, M.J. (1962), *Water Supply on Gibraltar*—American Water Works Association Journal.

Slatyer, R.O. and McIlroy I.C. (1961), *Practical Microclimatology.* UNESCO, Paris.

Spores, R. (1969), *Settlement, farming technology, and environment in the Nochixtlan Valley.* Science, 166 : 557-569.

Stone, E.C. (1957), *Dew as Ecological factor. A review of the Literature.* Ecology 38 : 407-413.

Twomey, S. (1957), *Precipitation by Direct Interception of Cloud Water.* Weather 12 : 120-122.

U.S. Public Health Service, (1965), *Manual of Individual Water Supply Systems.* USDHEW, Washington DC.

Vivian, G. (1974), Conservation and diversion-water control systems in the Anansazi Southwest; in T. Downing and G. McGuire, eds., *Irrigation's Impact on Society.* Anthropological Papers of the Univ. of Arizona 25. Tucson : U. Arizona Press.

Wagner, E.G. and Lanoix, J.N. (1959), Water Supply for Rural Areas and Small Communities.

Went, F.W. (1955), *Fog, Mist, Dew and Other Sources of Water.* Year Book of Agriculture. U.S. Dept. of Agriculture, Washington DC.

West, R. (1970), *Population densities and agricultural practices in pre-Columbian Mexico, with emphasis on semi-terracing.* Verhandlungen des XXXVIII Internationalen Amerikanisten-Kongresses, Stuttgart-München 2 : 361-369.

White, G. (1906), *The Natural History of Selborne,* London. Dent.

Wilken, G. (1976), *Studies of Resource Management in Traditional Middle American Farming Systems.* No. 5 : Surface Geometry, Part One: Slope Management. Colorado State Univ. Unpublished report.

Woodbury, R. (1961), *Prehistoric Agriculture at Point of Pines, Arizona.* Memoirs 17. Salt Lake City: Society for American Archaeology.

Woodbury, R. and Neely J. (1972), Water control systems of the Tehuacan Valley. In: F. Johnson ed. *The Prehistory of Tehuacan Valley,* Vol. 4, Chronology and Irrigation. Austin & London: University of Texas Press.

Yoder, E. J. (1959), *Principles of Pavement Design.*

RAIN AND STORM WATER HARVESTING

Experts' Group Meeting

30 October - 2 November 1979, Nairobi, Kenya

Participants

United Nations

Dr. Letitia E. Obeng	Chairman, Soil and Water Task Force UNEP, P.O. Box 30552, Nairobi
Dr. N. Gebremedhin	UNEP, Human Settlements Task Force P.O. Box 30552, Nairobi
Ms. Enid Burke	UNEP Information Division P.O. Box 30552, Nairobi
Ms. Abeba Wolderufael	UNICEF, Village Technology Unit P.O. Box 44145, Nairobi

Consultants

Dr. Kirsten Johnson	Department of Geography, Clark University, Worcester, Mass., 0610, USA
Dr. Liang Jian	Environment Protection Office of Ministry of Water and Conservancy and Electric Power, The People's Republic of China
Dr. Peter Schwerdtfeger	Professor of Meteorology, Flinders University of South Australia, Adelaide, Australia, 5042
Prof. Rama Prasad	Department of Civil Engineering, Indian Institute of Science, Bangalore 560 012, India

Dr. George Ongweny — Department of Geography, University of Nairobi, P.O. Box 30197, Nairobi

Dr. Georgy Kovacs — National Water Authority, 1394 Budapest, II. F.O. - UTCA 48-50 Hungary

Assessors

Dr. Ebbo H. Hofkes — WHO International Reference Centre for Community Water Supply, P.O. Box 140, Leidschendam, The Hague, Netherlands

Dr. Gerd-Jan de Kruijff — University of Nairobi, Housing Research and Development, P.O. Box 30197, Nairobi

Technical Editor

Dr. John J. Gaudet — University of Nairobi P.O. Box 30197 Nairobi, Kenya